NIMS Monographs

National Institute for
Materials Science

The NIMS Monographs are published by the National Institute for Materials Science (NIMS), a leading public research institute in materials science in Japan, in collaboration with Springer. The series present research results achieved by NIMS researchers through their studies on materials science as well as current scientific and technological trends in those research fields.

These monographs provide readers up-to-date and comprehensive knowledge about fundamental theories and principles of materials science as well as practical technological knowledge about materials synthesis and applications.

With their practical case studies the monographs in this series will be particularly useful to newcomers to the field of materials science and to scientists and engineers working in universities, industrial research laboratories, and public research institutes. These monographs will be also available for textbooks for graduate students.

National Institute for Materials Science
http://www.nims.go.jp/

More information about this series at http://www.springer.com/series/11599

Masayoshi Higuchi

Metallo-Supramolecular Polymers

Synthesis, Properties, and Device Applications

 Springer

Masayoshi Higuchi
National Institute for Materials Science
Tsukuba, Japan

ISSN 2197-8891   ISSN 2197-9502   (electronic)
NIMS Monographs
ISBN 978-4-431-56889-6   ISBN 978-4-431-56891-9   (eBook)
https://doi.org/10.1007/978-4-431-56891-9

This Springer imprint is published by the registered company Springer Japan KK part of Springer Nature.
The registered company address is: Shiroyama Trust Tower, 4-3-1 Toranomon, Minato-ku, Tokyo 105-6005, Japan

Preface

Metallo-Supramolecular polymers are a new type of polymer, in which the polymer chains are formed by the complexation of metal ions with multi-topic organic ligands. Therefore, the polymer chains are composed of coordinate covalent bonds. In contrast, conventional organic polymers such as polypropylene are prepared by polymerization of the monomers via the formation of covalent bonds. This difference in the polymer backbones between metallo-supramolecular polymers and conventional organic polymers allows large structural and functional differences in the polymers. From the electronic interaction between the metal ion and the ligand, unique electrochemical, optical, magnetic, and catalytic properties are expected in metallo-supramolecular polymers. This book is written so that non-experts in this research field can understand polymer design, synthetic methodologies, electrochromic/ionic/emissive properties, and display device applications of metallo-supramolecular polymers. In this book, Fe(II)-, Ru(II)-, Co(II)-, Zn(II)-, Cu(II)-, Pt(II)-, Ni(II)-, Cd(II)-, Mo(VI)-, and Eu(III)-based metallo-supramolecular polymers are introduced as examples. The polymer design, synthetic methods, and electrochemical and optical properties are included in each chapter. The polymer design section contains not only linear polymers but hyperbranched polymers. Heterometallo-Supramolecular polymers with two metal ion species are also presented. Electrochromism (blue, red, yellow, black, and multi-color), emission, ionic/protonic conduction, nonvolatile memory, vapoluminescence, and electrochemical switching of emission are described as representative properties. Metallo-Supramolecular polymer films are amorphous and suitable for device applications. This book discusses (solid-state and flexible) electrochromic display devices and humidity sensors. The polymers described in this book are limited examples from the expected huge research field of metallo-supramolecular polymer chemistry, because numerous combinations of metal ions and organic ligands in polymer synthesis will create novel metallo-supramolecular polymers, whose functions are unimaginable now, in the future.

Tsukuba, Japan   Masayoshi Higuchi

Contents

Chapter 1
Introduction

1.1 Scope of This Book

Metallo-Supramolecular polymers are a new kind of polymeric compound. There-fore, a basic understanding of conventional organic polymers is required before the introduction of metallo-supramolecular polymers. Sections 1.2–1.5 of this chapter describe the basic concepts of conventional organic polymers. Sections 1.6–1.9 present the fundamental characteristics of metallo-supramolecular polymers in comparison with organic polymers.

Chapters 2–11 present design, synthesis, electrochromic/ionic/emissive prop-erties, and display device applications of metallo-supramolecular polymers. Each chapter focuses on one metal species. Ten metal species (Fe, Ru, Co, Zn, Cu, Pt, Ni, Cd, Mo, and Eu) in total are introduced as the metal included in the polymer in this book (Fig. 1.1). When you read these chapters, you will notice that the incorpo-rated metal species changes the properties of the metallo-supramolecular polymer considerably. In addition, you will find that the design of the organic ligands is quite important to obtain polymers with the desired properties. On the basis of the amor-phous nature of metallo-supramolecular polymers, the polymer films are obtained by general film-preparation methods including spin-coating and spray-coating. This book also discusses the fabrication of solid-state devices with the polymer films, including electrochromic display devices, and the properties of the devices.

1.2 History of Polymer

The fundamental concept of "macromolecule" was firstly proposed by Hermann Staudinger, who was awarded the Nobel Prize in Chemistry in 1953. In general, "macromolecule" is defined as a molecule with a high molecular weight of more than 10,000. "Polymer" is another word used to represent such a large molecule, but the meanings of the two words are slightly different. The former simply means huge molecules, but the later also includes the concept of the repetition of molecular units, because "polymer" is an extension of the words "monomer", "dimer", "trimer", and

© National Institute for Materials Science, Japan 2019
M. Higuchi, *Metallo-Supramolecular Polymers*, NIMS Monographs,
https://doi.org/10.1007/978-4-431-56891-9_1

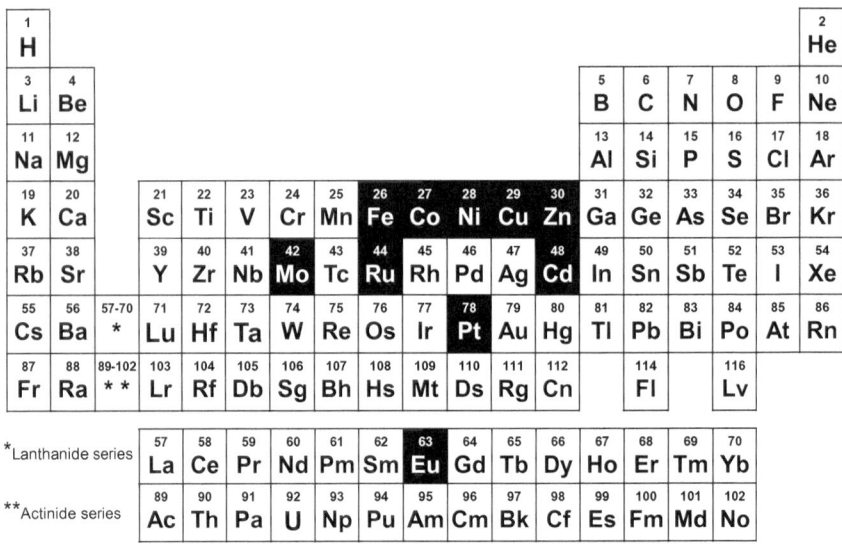

Fig. 1.1 Ten metal species (Fe, Ru, Co, Zn, Cu, Pt, Ni, Cd, Mo, and Eu) incorporated into the metallo-supramolecular polymers (the elements are shown in white on black backgrounds in the periodic table)

"tetramer" (Fig. 1.2a). Since Staudinger had suggested "macromolecule", macro-molecular (or polymer) science has been extensively developed, and polymeric materials such as plastics and rubbers have made our daily lives more convenient

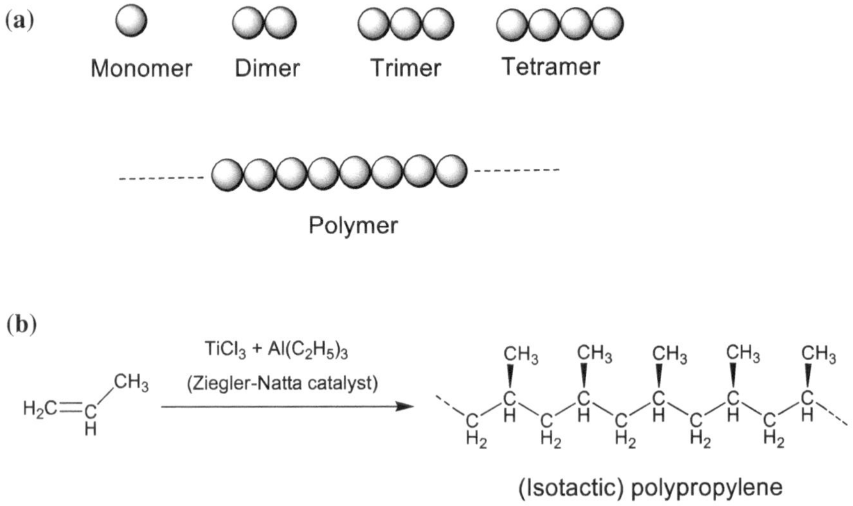

Fig. 1.2 a Polymeric molecular structures. **b** Synthesis of polypropylene using Ziegler–Natta catalyst

than before their appearance. For example, plastic bottles made of polyethylene terephthalate (PET) are lighter and tougher than metal or glass bottles.

Another epoch-making event in the history of polymer science is the development of polymer synthetic methods. The development of polymer synthesis started with the preparation of polyethylene and polypropylene using Ziegler–Natta catalysts (Fig. 1.2b). Karl Ziegler and Giulio Natta were also awarded the Nobel Prize in Chemistry in 1963 for their work. The starting compound to be polymerized is called the "monomer". A linear structure of polymers is obtained by the polymerization of monomers bearing two connecting sites. The bonds connecting monomers in polymers are usually covalent bonds. Against decomposition, covalent bonds are much stronger than other bonds such as coordinate covalent bonds and hydrogen bonds because covalent bond formation, especially between carbon atoms, is an energetically very favorable reaction, whereas the formation of coordinate covalent and hydrogen bonds is much less so.

1.3 Linear Polymer Structures

Polycondensation is a representative polymerization method and describes a condensation reaction between monomers to produce polymers. For example, polyamides are prepared by the polycondensation of amines with carboxylic acids. Polyesters are synthesized by the polycondensation of alcohols with carboxylic acids. In polymer synthesis by polycondensation, a difference in a monomer structure (an AB monomer vs. A_2/B_2 monomers) changes the corresponding polymer structure (Fig. 1.3a). When a monomer has two different reaction sites, it is called an "AB monomer". The polycondensation of AB monomers results in the formation of polymers with "a head-to-tail structure" due to the intermolecular connection between the different reaction sites. On the other hand, polymers with "a head-to-head, tail-to-tail structure" are obtained by the polycondensation of A_2 monomers and B_2 monomers, which have two reaction sites that are the same. For instance, head-to-tail poly(α-phenyl)phenylazomethine is formed by the polycondensation of the AB monomer bearing one amino and one carboxyl site (Fig. 1.3b) [1, 2]. A head-to-head, tail-to-tail polymer is synthesized by the polycondensation of phenylene diamine (an A_2 monomer) and phenylene dicarbonyl (a B_2 monomer) [3].

1.4 Hyperbranched Polymer Structures

When a monomer has more than three binding sites, various polymer structures including hyperbranched, dendritic, ladder, and networking structures can be formed (Fig. 1.4a). Hyperbranched polymers are polymers with a highly branched structure.

(a)

AB monomer   Head-to-tail polymer

A$_2$ / B$_2$ monomers   Head-to-head, tail-to-tail polymer

(b)

Fig. 1.3 a A head-to-tail polymer formed by polycondensation of an AB monomer and a head-to-head, tail-to-tail polymer formed using A$_2$/B$_2$ monomers. **b** Synthesis of head-to-tail and head-to-head, tail-to-tail poly(α-phenyl)phenylazomethines

This means that hyperbranched polymers also have many terminal units. For example, the polymerization of AB$_2$ monomers or A$_2$/B$_3$ monomers often gives a hyperbranched structure (Fig. 1.4b). The structures of the two polymers are slightly different. In the polymerization of AB$_2$ monomers, the high-molecular-weight polymers have mostly only site B as the terminus. In the polymerization of A$_2$/B$_3$ monomers, the total numbers of A and B sites in the termini are the same when the A$_2$ and B$_3$ monomers are mixed at a 3:2 molar ratio. Intramolecular reactions in the hyperbranched polymer result in a network polymer formation. Dendrimers are a special structure in hyperbranched polymers: a regularly repeated structure of branches from the central unit (core) to the periphery [4]. The number of repetitions is called the number of "generations". The overall shape of high-generation dendrimers is spherical [5–12].

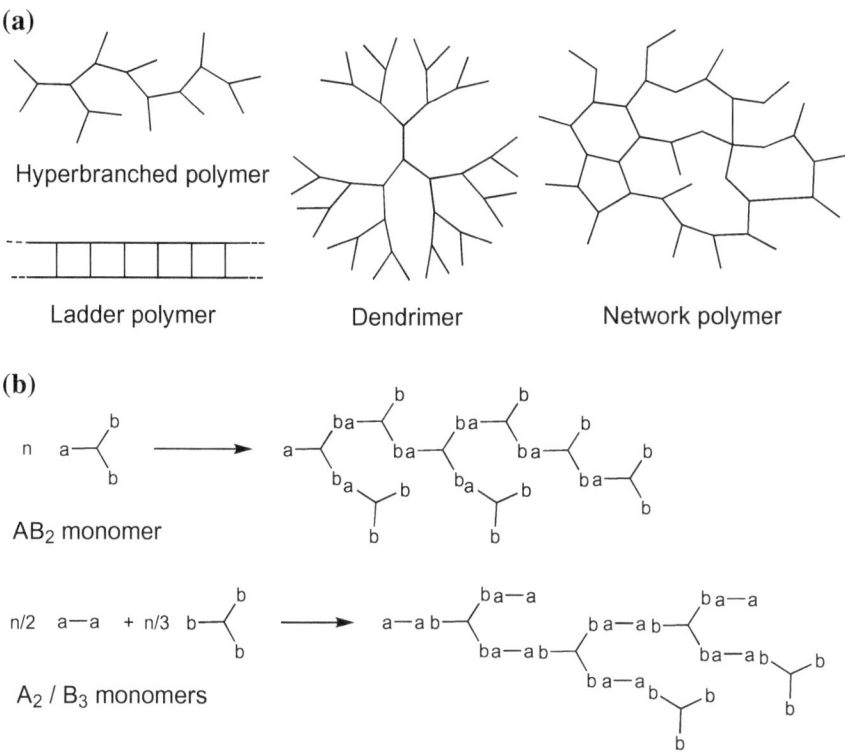

Fig. 1.4 **a** Various polymer structures: hyperbranched polymer, ladder polymer, dendrimer, and network polymer. **b** A head-to-tail polymer formed by polycondensation of an AB$_2$ monomer and a head-to-head, tail-to-tail polymer formed from A$_2$/B$_3$ monomers

1.5   π-Conjugated Polymers

In general, organic polymers are electronic insulators, because no free electrons exist in the materials, unlike metals. Shirakawa et al. accidentally found electronic conductivity in polyacetylene (Fig. 1.5a) [13]. Alan J. Heeger, Alan G. MacDiarmid, and Hideki Shirakawa were awarded the Nobel Prize in Chemistry in 2000 for their development of electrically conductive polymers. To realize electronic conduction in polymers, (1) a π-conjugated polymer structure and (2) chemical or electrochemical doping are required. Doping means the chemical/electrochemical removal or addition of electrons from or to the polymer. The positive or negative charges formed by doping serve as charge carriers through the polymer chain. Polymers bearing alternating double and single bonds are called π-conjugated polymers [14, 15]. The large overlap of neighboring π-orbitals enhances the π-conjugation and decreases the bandgap between the highest occupied molecular orbital (HOMO) and the lowest unoccupied molecular orbital (LUMO). Figure 1.5b shows the band structures in insulators, semiconductors, and metals. The large bandgap between the valence band and the

(a)

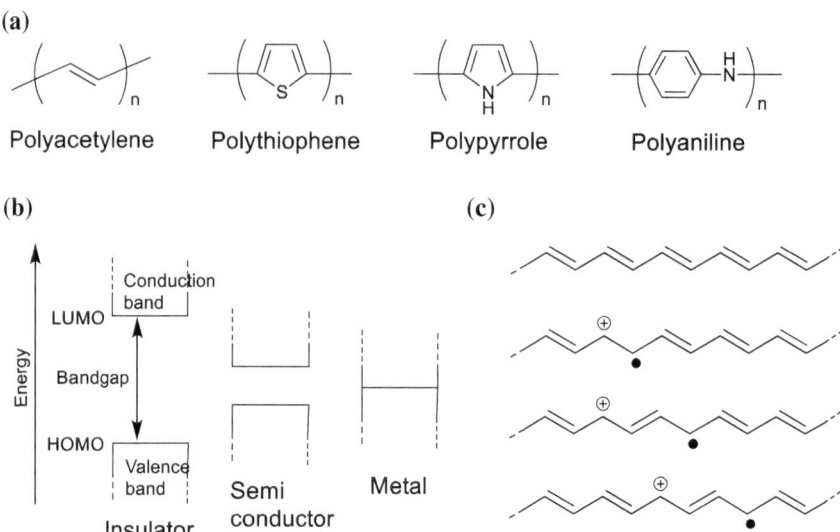

Polyacetylene   Polythiophene   Polypyrrole   Polyaniline

(b)   **(c)**

Fig. 1.5 **a** Representative π-conjugated polymers. **b** Energy diagrams for insulators, semi-conductors, and metals. **c** A schematic presentation of movement of a cation and a radical through a polyacetylene chain

conduction band is the reason for the poor electrical conductivity in an insulator, and the absence of a bandgap between the two bands is the reason for the high conductivity in metals. The π-conjugated polymers are categorized as semiconducting materials. Doping in π-conjugated polymers causes a new band between the valence band and the conduction band and makes the polymer conductive. The movement of the charges generated by doping through the polymer chain is presented in Fig. 1.5c.

1.6   Metallo-Supramolecular Polymers

Supramolecular chemistry has rapidly developed over the last 40 years. Supramolecules are compounds in which more than two molecules are assembled by noncovalent bonds (hydrogen bonds, π-π interactions, van der Waals interaction, etc.) and coordinate covalent bonds. Donald J. Cram, Jean-Marie Lehn, and Charles J. Pedersen are the pioneers of supramolecular chemistry and were awarded the Nobel Prize in Chemistry in 1987 for the development and use of molecules with structure-specific interactions of high selectivity [16].

Supramolecules with a polymeric structure are called supramolecular polymers. When the polymer includes metal species, it is called a metallo-supramolecular polymer. Metallo-Supramolecular polymers are a new category of polymer [17], because the polymer backbone is composed of coordinate covalent bonds, unlike conventional organic polymers with carbon–carbon covalent bonds. They are synthesized

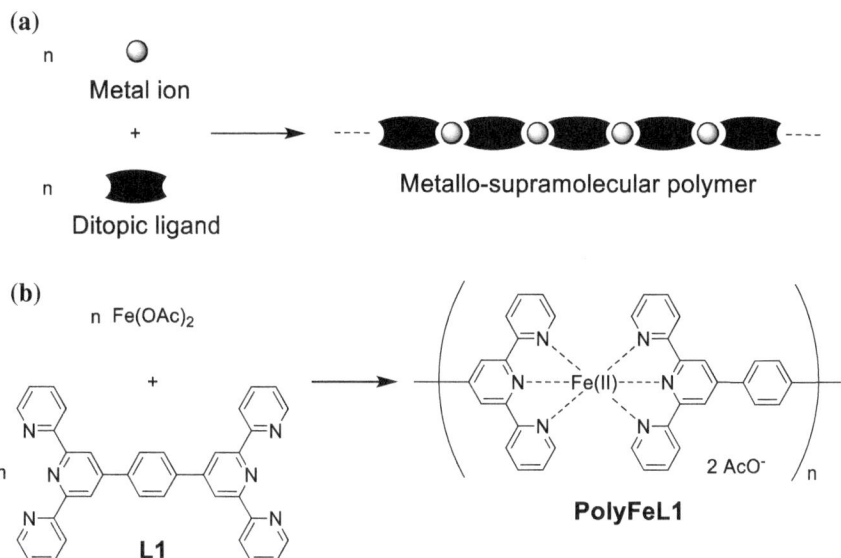

Fig. 1.6 **a** Metallo-Supramolecular polymer formation by the 1:1 complexation of metal ions with ditopic organic ligands. **b** Fe(II)-based metallo-supramolecular polymer (**polyFeL1**) synthesis using Fe(II) ions and 1,4-bis(2,2':6',2''-terpyridin-4'-yl)benzene (**L1**)

by the 1:1 complexation of metal ions and ditopic organic ligands (Fig. 1.6a) [18–23]. Ditopic ligands are compounds bearing two metal-coordination sites. For example, Fe(II) ions are complexed with 1,4-bis(2,2':6',2''-terpyridin-4'-yl)benzene (**L1**) to form the corresponding Fe(II)-based metallo-supramolecular polymer (**polyFeL1**) with an Fe(II) ion and **L1** introduced alternately (Fig. 1.6b). When $Fe(OAc)_2$ is used as the metal salt for the polymerization, acetate anions are also included in the polymer to neutralize the positive charge of the Fe(II) ions.

1.7 Degree of Polymerization

Molecular weight and degree of polymerization are unique and useful parameters for characterizing organic polymers because the covalent bonds in the polymers are not easily broken after they are formed. However, unlike conventional organic polymers, metallo-supramolecular polymers change the polymer length in solution because the formation and decomposition of coordinate covalent bonds are reversible based on the equilibrium reactions. Therefore, the complexation constant (K) in the coordinate covalent bond formation is an important parameter to decide the degree of polymerization of the polymer. High complexation constants are desirable to obtain long polymer chains. Multi-dentate coordination sites in organic ligands such as terpyridine have stronger binding affinity to metal ions than mono-dentate sites such

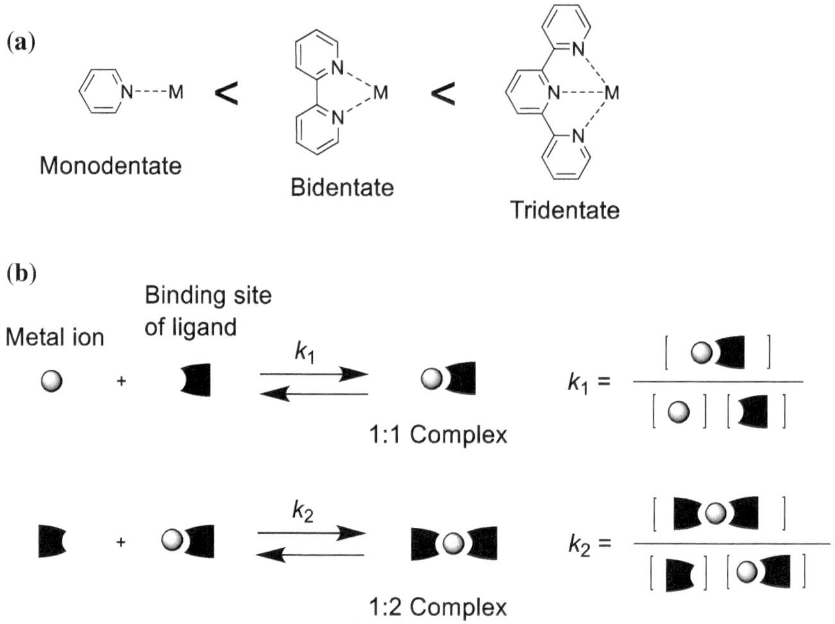

(a)

Monodentate

Bidentate

Tridentate

(b)

Binding site
of ligand

Metal ion

1:1 Complex

$$k_1 = \frac{[\;\text{O}\blacktriangleleft\;]}{[\text{O}\,]\,[\blacktriangleleft\,]}$$

1:2 Complex

$$k_2 = \frac{[\;\blacktriangleright\text{O}\blacktriangleleft\;]}{[\blacktriangleright\,]\,[\text{O}\blacktriangleleft\,]}$$

Fig. 1.7 **a** Different stabilities of metal complexes among monodentate, bidentate, and tridentate ligands. **b** Equilibrium in the complexation between a metal ion and a binding site of ligand and the equations to determine the complexation constants (k_1 and k_2)

as pyridine (Fig. 1.7a) since multi-dentate coordination stabilizes the metal complex more strongly than mono-dentate coordination. One must be careful of the different complexation constants between the first coordination and the second coordination of the metal complex (Fig. 1.7b). The complexation constant (k_2) for the second complexation is often lower than that (k_1) for the first complexation, because the metal ion in the complex formed by the first complexation has a lower cationic character than that before the complexation. In addition, the second complexation is sometimes prevented due to the steric hindrance of the coordinated site in the metal complex formed by the first coordination.

1.8   Absorptions of Metal Complexes

Conventional organic polymers such as polypropylene have no color because they have no absorptions in the visible region. To absorb in the visible region, the polymer must have an electronic transition with a bandgap energy less than 3.26 eV (380 nm). π-Conjugated polymers show unique electronic and optical properties, such as conductivity and fluorescence. These properties are mainly caused by π-π* electronic transition in the aromatic moieties. The bandgap energy of π-π* transition decreases

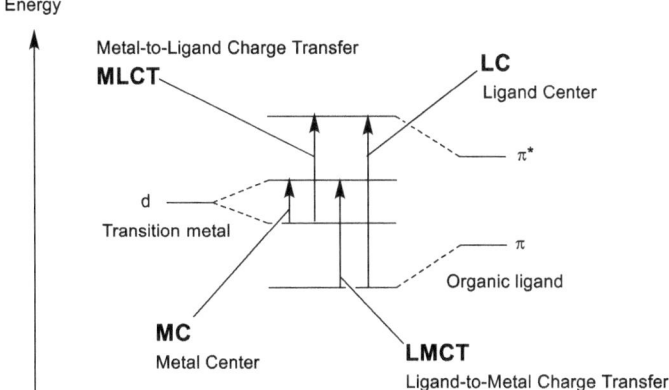

Fig. 1.8 Electronic transitions in transition metal complexes

when the π-conjugation is extended. Therefore, long π-conjugated polymers have a color.

Metallo-Supramolecular polymers, on the other hand, have several electronic transitions based on the metal complex moieties. In general, metal complexes exhibit metal center (MC) electronic transitions such as d-d* transitions in transition metal complexes and d-f* transitions in lanthanide metal complexes. When the organic ligand includes a π-conjugated structure, various transitions are expected owing to the combinations of the d-d* transition (metal center (MC) transition) of the metal and the π-π* transition (ligand center (LC) transition) of the ligand (Fig. 1.8). Electronic transitions from d orbitals of the transition metal to π* orbitals of the ligand are called metal-to-ligand charge transfer (MLCT). Similarly, electronic transitions from π orbitals of the ligand to d* orbitals of the transition metal are ligand-to-metal charge transfer (LMCT). Therefore, metallo-supramolecular polymers are expected to have various electronic and optical properties based on these transitions. In addition, the polymer films are easily formed on an electrode by general methods of film preparation such as spin-coating. Therefore, the redox activity of the metal ions in the polymer film can be caused by an applied electrochemical potential. Since changes in the oxidation state of the metal ions vary the band structure of the metal complex in the polymer, the electronic and optical properties of the polymers are expected to be controlled electrochemically.

1.9 Basic Design of Metallo-Supramolecular Polymers

Metallo-Supramolecular polymers with a linear structure are synthesized by the 1:1 complexation of metal salts with ditopic organic ligands (Fig. 1.9). The total number of coordination sites in the ligand should be the same as the coordination number of the metal ion. For example, terpyridine is a tridentate ligand with three coordination

Fig. 1.9 A general structure of a linear metallo-supramolecular polymer (M: metal ion; Spacer: a spacer unit between the two terpyridines; R: a substituent)

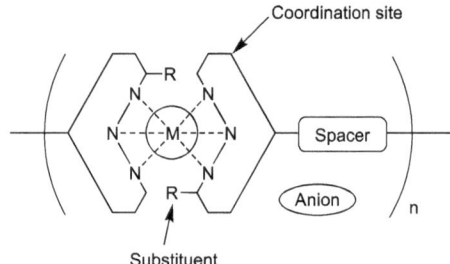

sites. Therefore, bis(terpyridine)s have six coordination sites in total. Since Fe(II), Ru(II), Co(II), and Zn(II) ions have an octahedral geometry with a coordination number of six, bis(terpyridine)s are ideal ditopic ligands for metallo-supramolecular formation with these metal ions. If the numbers are different between the ligand and the metal, the complexation results in branched polymer formation. The introduction of substituents (R in Fig. 1.9) to the coordination site on the ligand changes the properties of the polymers due to the steric hindrance and/or electron donating/withdrawing effects of the substituents. Counter anions of the metal salts are incorporated into the polymers to neutralize the positive charge of the metal ions. The counter anion in the polymer affects the polymer properties such as solubility and film-forming ability. Exchanging the counter anions in the polymer to other anions is possible by immersing the polymer in a solution that contains a large amount of other anions. A spacer unit connecting two coordination sites in the ligand is also important in determining polymer properties. For instance, Rowan et al. reported multi-stimuli, multi-responsive metallo-supramolecular polymers. Since their polymers had a flexible aliphatic spacer, the macrocyclization of the polymer chain occurred owing to the flexibility of the polymer backbone. When a rigid aromatic unit was used as the spacer, the polymer structure was also rigid and linear. The oxidation state of the metal ions in the polymer was controlled chemically or electrochemically. The chemical control was achieved by the addition of an oxidant or a reductant to a solution of the polymer. The electrochemical control was performed in a polymer film prepared on an electrode. In the following chapters, the polymer design, synthetic methods, electrochemical/ionic/emissive properties, and the applications of metallo-supramolecular polymers to devices are introduced using Fe(II)-, Ru(II)-, Co(II)-, Zn(II)-, Cu(II)-, Pt(II)-, Ni(II)-, Cd(II)-, Mo(VI)-, and Eu(III)-based polymers as examples.

References

1. Higuchi M, Yamamoto K (1999) Novel cyclic molecules: selective synthesis of cyclic phenylazomethines. Org Lett 1:1881–1883. https://doi.org/10.1021/ol990975s
2. Higuchi M, Kimoto A, Shiki S, Yamamoto K (2000) Selective synthesis of novel cyclic phenylazomethine trimers. J Org Chem 65:5680–5684. https://doi.org/10.1021/jo000509c

3. Higuchi M, Kanazawa H, Yamamoto K (2003) The first electroresponsive phenylazomethine macrocycles: highly preferential formation and regular molecular packing. Org Lett 5:345–347. https://doi.org/10.1021/ol0273755

4. Tomalia DA, Baker H, Dewald J, Hall M, Kallos G, Martin S, Roeck J, Ryder J, Smith P (1985) A new class of polymers: starburst-dendritic macromolecules. Polym J 17:117–132. https://doi.org/10.1295/polymj.17.117

5. Higuchi M, Shiki S, Yamamoto K (2000) Novel phenylazomethine dendrimers: synthesis and structural properties. Org Lett 2:3079–3082. https://doi.org/10.1021/ol006241t

6. Higuchi M, Shiki S, Ariga K, Yamamoto K (2001) First synthesis of phenylazomethine dendrimer ligands and structural studies. J Am Chem Soc 123:4414–4420. https://doi.org/10.1021/ja004239r

7. Higuchi M, Kanazawa H, Yamamoto K (2001) Controlled cyclotrimerization in hyperbranched polymer synthesis. Macromolecules 34:8847–8850. https://doi.org/10.1021/ma0111461

8. Yamamoto K, Higuchi M, Shiki S, Tsuruta M, Chiba H (2002) Stepwise radial complexation of imine groups in phenylazomethine dendrimers. Nature 415:509–511. https://doi.org/10.1038/415509a

9. Higuchi M, Tsuruta M, Chiba H, Shiki S, Yamamoto K (2003) Control of stepwise radial complexation in dendritic polyphenylazomethines. J Am Chem Soc 125:9988–9997. https://doi.org/10.1021/ja035608x

10. Satoh N, Cho JS, Higuchi M, Yamamoto K (2003) Novel triarylamine dendrimers as a hole-transport material with a controlled metal-assembling function. J Am Chem Soc 125:8104–8105. https://doi.org/10.1021/ja034811p

11. Nakajima R, Tsuruta M, Higuchi M, Yamamoto K (2004) Fine control of the release and encapsulation of Fe ions in dendrimers through ferritin-like redox switching. J Am Chem Soc 126:1630–1631. https://doi.org/10.1021/ja037480p

12. Takanashi K, Chiba H, Higuchi M, Yamamoto K (2004) Efficient synthesis of poly(phenylazomethine) dendrons allowing access to higher generation dendrimers. Org Lett 6:1709–1712. https://doi.org/10.1021/ol049656d

13. Shirakawa H, Louis EJ, MacDiarmid AG, Chang CK, Heeger AJ (1977) Synthesis of electrically conducting organic polymers: halogen derivatives of polyacetylene, $(CH)_x$. J Chem Soc-Chem Comm 578–580. https://doi.org/10.1039/c39770000578

14. Higuchi M, Imoda D, Hirao T (1996) Redox behavior of polyaniline-transition metal complexes in solution. Macromolecules 29:8277–8279. https://doi.org/10.1021/ma960761f

15. Higuchi M, Ikeda I, Hirao T (1997) A novel synthetic metal catalytic system. J Org Chem 62:1072–1078. https://doi.org/10.1021/jo9617575

16. Lehn JM (1988) Supramolecular chemistry-scope and perspectives molecules, supermolecules, and molecular devices (Nobel lecture). Angew Chem Int Ed Engl 27:89–112. https://doi.org/10.1002/anie.198800891

17. Schubert US, Hien O, Eschbaumer C (2000) Functionalized polymers with metal complexing segments: a simple and high-yield entry towards 2,2′:6′,6″-terpyridine-based oligomers. Macromol Rapid Commun 21:1156–1161. https://doi.org/10.1002/1521-3927(20001101)21:16%3c1156:aid-marc1156%3e3.0.co;2-o

18. Kolb U, Buscher K, Helm CA, Lindner A, Thunemann AF, Menzel M, Higuchi M, Kurth DG (2006) The solid-state architecture of a metallosupramolecular polyelectrolyte. Proc Natl Acad Sci USA 103:10202–10206. https://doi.org/10.1073/pnas.0601092103

19. Kurth DG, Higuchi M (2006) Transition metal ions: weak links for strong polymers. Soft Matter 2:915–927. https://doi.org/10.1039/b607485e

20. Pal RR, Higuchi M, Kurth DG (2009) Optically active metallo-supramolecular polymers derived from chiral bis-terpyridines. Org Lett 11:3562–3565. https://doi.org/10.1021/ol901293r

21. Pal RR, Higuchi M, Negishi Y, Tsukuda T, Kurth DG (2010) Fluorescent Fe(II) metallo-supramolecular polymers: metal-ion-directed self-assembly of new bisterpyridines containing triethylene glycol chains. Polym J 42:336–341. https://doi.org/10.1038/pj.2010.3

22. Li J, Futera Z, Li H, Tateyama Y, Higuchi M (2011) Conjugation of organic-metallic hybrid polymers and calf-thymus DNA. Phys Chem Chem Phys 13:4839–4841. https://doi.org/10.1039/c0cp02037k

23. Li J, Murakami T, Higuchi M (2013) Metallo-supramolecular polymers: versatile DNA binding and their cytotoxicity. J Inorg Organomet Polym Mater 23:119–125. https://doi.org/10.1007/s10904-012-9752-2

Chapter 2
Fe(II)-Based Metallo-Supramolecular Polymers

2.1 Synthesis

Fe(II)-based metallo-supramolecular polymers (**polyFeL1-5**) (Fig. 2.1a) are prepared using the 1:1 complexation of Fe(II) acetate with **L1-5** [1–4].

Fe(II) ions have an octahedral geometry with six coordination sites. It is known that a stable 1:2 complex of the metal ion with tridentate ligands such as terpyridines is formed by the complexation. Therefore, metallo-supramolecular polymer formation is expected by the 1:1 complexation of Fe(II) ions with bis(terpyridine)s bearing two terpyridine moieties. Bis(terpyridine)s **L1-5** are designed for the synthesis of Fe(II)-based metallo-supramolecular polymers (**polyFeL1-5**).

The two-step Kröhnke procedure has been used to prepare terpyridine rings: an aldol condensation of benzaldehydes and 2-acetylpyridines, followed by a Michael addition of the resultant azachalcone with pyridinium iodide. A Suzuki-type cross-coupling may be used to synthesize **L2** and **L4**. 1,4-Bis(2,2′:6′,2″-terpyridine-4′-yl)benzene (**L1**) is commercially available from Aldrich company. As an example of the ligand synthesis, the detailed synthetic procedure for **L4** (Fig. 2.1b) is described as follows. Four steps are necessary to obtain the ligand.

(i). To a methanol solution (100 mL) of 4-bromobenzaldehyde (1.0 g, 5.40 mmol) is added 2-acetyl-6-bromopyridine (1.1 g, 5.40 mmol) and 2% aqueous NaOH (0.22 g, 11 mL). The mixture is stirred for 2 h at room temperature. The precipitate is collected by filtration and washed sequentially with water and methanol three times. The slight yellow solid is dried in vacuo to give the product, 1-(3-oxo-3-[2-(6-bromopyridyl)]propen-1-yl)-4-bromobenzene (1.41 g, 71%), which is used directly in the following reaction without further purification.

(ii). To a methanol solution (100 mL) of the product (2.50 g, 6.81 mmol) is added 1-acetophenonepyridinium iodide (2.22 g, 6.81 mmol) and NH_4OAc (15.73 g, 204.3 mmol). The solution is refluxed for 24 h, and then cooled to room temperature. The precipitate is collected by filtration and washed sequentially with water and methanol three times. The precipitate is purified by column

© National Institute for Materials Science, Japan 2019
M. Higuchi, *Metallo-Supramolecular Polymers*, NIMS Monographs,
https://doi.org/10.1007/978-4-431-56891-9_2

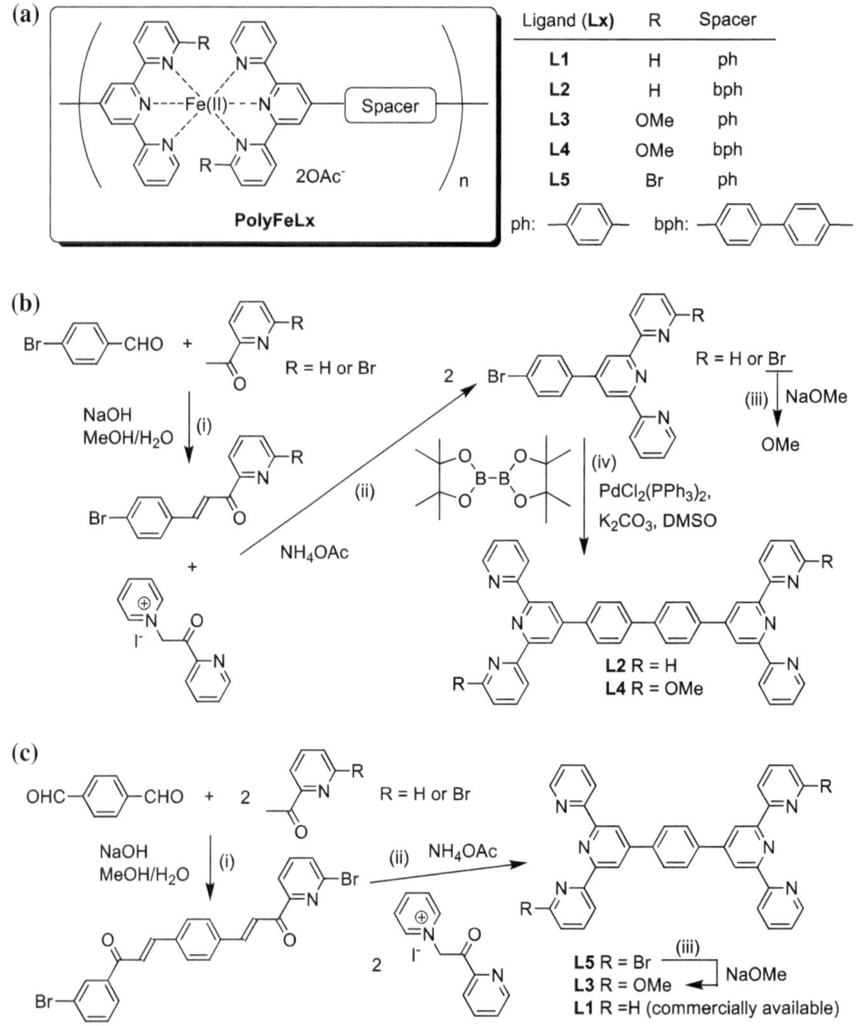

Fig. 2.1 **a** Fe(II)-based metallo-supramolecular polymers (**polyFeL1-5**) and the synthesis of **b L2** and **L4** and **c L3** and **L5**

       chromatography on silica gel (methylene chloride/hexane 2:1) to give pure
       4′-(4-bromophenyl)-6-bromo-2,2′:6′,2″-terpyridine (2.36 g, 74%) as a white
       powder.

(iii). To a dioxane solution (60 mL) of the product (800 mg, 1.71 mmol) is added
       NaOMe (1.0 M solution in methanol, 5.14 mmol). The solution is stirred at
       100 °C for 2 days under nitrogen atmosphere. The solvent is evaporated under
       reduced pressure and the residue is suspended in chloroform (150 mL). The
       suspension is then washed with water (50 mL) three times. The organic layer

is separated and dried over MgSO$_4$, then filtered, concentrated, and purified by column chromatography on activated basic Al$_2$O$_3$ (hexane/methylene chloride = 1:1) to give 4′-(4-bromophenyl)-6-methoxyl-2,2′:6′,2″-terpyridine (688 mg, 96%) as a white powder.

(iv). To a DMSO solution (20 mL) of the product (100 mg, 0.24 mmol) is sequentially added bis(pinacolato)diboron (31 mg, 0.12 mmol), K$_2$CO$_3$ (99 mg, 0.72 mmol), and PdCl$_2$(PPh$_3$)$_2$ (8.5 mg, 5.0 %mmol). The solution is degassed and stirred at 80 to 100 °C under an argon atmosphere until the starting material disappears as monitored by TLC. The catalyst is removed by filtration and washed thoroughly with chloroform after the reaction mixture is cooled to room temperature. The filtrate is then washed with water (ca. 50 mL) five times. The organic layer is separated, dried over MgSO$_4$, filtered, concentrated, and purified by column chromatography on activated basic Al$_2$O$_3$ (methylene chloride/hexane = 4:1, then pure methylene chloride) to give **L4** (62 mg, 76%) as a white powder.

As another example of ligand synthesis, the detailed synthetic procedure for **L3** (Fig. 2.1c) is carried out as follows. Three steps are necessary to obtain the ligand.

(i). To a methanol solution (200 mL) of benzene-1,4-dicarboxaldehyde (3.62 g, 27.0 mmol) is added 2-acetyl-6-bromopyridine (10.80 g, 53.98 mmol) and 10% aqueous NaOH (2.16 g in 20 mL of water). The mixture is stirred at room temperature overnight. The precipitate is collected by filtration and washed sequentially with water and methanol three times. The slightly yellow solid is dried in vacuo to give diazachalcone (10.5 g, 78%), which is used directly for the following reaction without further purification.

(ii). To an ethanol solution (200 mL) of the product (7.1 g, 14 mmol) is added 1-acetophenonepyridinium iodide (9.3 g, 29 mmol) and NH$_4$OAc (54 g, 0.70 mol). The solution is refluxed for 60 h and then cooled to room temperature. The precipitate is collected by filtration and washed sequentially with water and methanol three times. The crude product is purified by column chromatography on basic Al$_2$O$_3$ (methylene chloride/hexane = 5:1 followed by pure methylene chloride) to give 1,4-bis-(6-bromo-2,2′:6′,2″-terpyridine-4-yl)benzene (**L5**) (3.9 g, 39%) as a white powder.

(iii). To a dioxane solution (60 mL) of **L5** (200 mg, 0.287 mmol) is added NaOMe (1.0 M solution in methanol, 1.72 mmol). The solution is refluxed for 2 days under nitrogen atmosphere. The solvent is evaporated under reduced pressure, and the residue is suspended in chloroform (200 mL). The suspension is then washed with water (50 mL) three times. The organic layer is separated, dried over MgSO$_4$, filtered, concentrated, and purified by column chromatography on activated basic Al$_2$O$_3$ (hexane/methylene chloride = 1:9, then pure methylene chloride) to give 1,4-bis(6-methoxyl-2,2′:6′,2″-terpyridine-4-yl)benzene (**L3**) (161 mg, 96%) as a white powder.

Fe(II)-based metallo-supramolecular polymers (**polyFeL1-5**) are synthesized by the 1:1 complexation of Fe(II) acetate with **L1-5**. The detailed synthetic procedure is as follows. Equimolar amounts of $Fe(OAc)_2$ and the ligand are refluxed in argon-saturated acetic acid (ca. 10 mL of solvent per mg of ligand) for 24 h. The solution is cooled to room temperature and filtered to remove a small amount of insoluble residues. The filtrate is moved to a petri dish and the solvent is evaporated slowly to dryness. The brittle film is collected and dried further in vacuo overnight to give the corresponding polymers (yield: >90%). The complexation of Fe(II) ions with bis(terpyridine)s during the polymer synthesis is confirmed by a color change to blue in the reaction mixture, because the complexes have the MLCT absorption bands. Although the ligands are hydrophobic, the polymers are soluble in polar solvents such as water and methanol owing to the hydrophilic metal complex moieties. Molecular weight of the polymers couldn't be determined by GPC measurement because of the decomplexation in the GPC column.

2.2   Electrochemical and Optical Properties

The optical properties of **polyFeL1-5** in methanol solution are measured by UV-vis spectroscopy (Table 2.1). The MLCT absorption from the Fe(II) ion to the terpyridine moiety in the polymers is observed at around 580 nm. The maximum wavelength (λ_{max}) and absorption coefficient (ε) for the MLCT absorption are different for different polymers. The introduction of substituents (electron-donating or electron-releasing groups) to the terpyridine moieties and the change in the spacer moiety between the two terpyridines in the ligands are considered to change the LUMO potential of the terpyridine moiety and/or distort the octahedral structure (Fig. 2.2a). The MLCT absorption is blue-shifted by the introduction of a biphenyl unit as a spacer on the ligand, probably because the steric hindrance between hydrogens in the biphenyl unit results in the decrease of π-conjugation and the increase of π^* potential (Fig. 2.2b). The introduction of methoxy groups as electron-releasing groups on the ligand causes a large decrease in ε without changing λ_{max}. The introduction of

Table 2.1 Optical and electrochemical properties of Fe(II)-based metallo-supramolecular polymers (**polyFeL1-5**)

	Maximum wavelength (λ_{max})[a] (nm)	Absorption coefficient (ε)[a], $\times 10^4$	Redox potential ($E_{1/2}$)[b], V versus Ag/Ag^+
PolyFeL1	585	3.03	0.77
PolyFeL2	579	2.57	0.78
PolyFeL3	585	1.43	0.70
PolyFeL4	578	1.67	0.70
PolyFeL5	612	0.77	0.93

[a]Solvent: methanol. [b]Working electrode: glassy carbon; counter electrode: Pt wire; reference electrode: Ag/Ag^+; electrolyte: 0.10 M n-Bu_4NClO_4/acetonitrile; scan rate: 100 mV/s

(a)

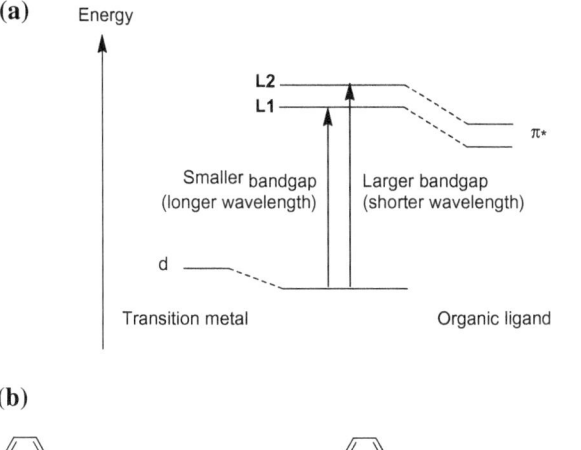

(b)

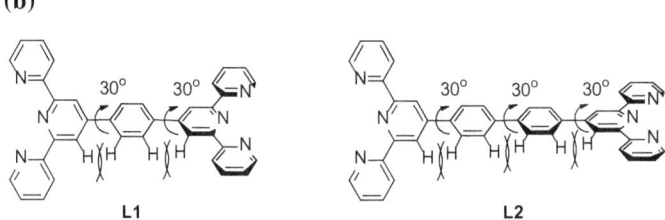

Fig. 2.2 a Energy diagram of the MLCT absorption in **polyFeL1** and **polyFeL2**. **b** Molecular geometries of **L1** and **L2**, which are suggested from X-ray crystal analysis of **L2**

bromo groups as electron-withdrawing groups results in a large decrease in ε and a red-shift in the λ_{max}.

The electrochemical properties of **polyFeL1-5** are measured using cyclic voltammetry (CV) (Table 2.1). For the measurement, a thin film of the polymer is prepared on a glassy carbon electrode and used as a working electrode. A reversible redox (reduction and oxidation) wave is observed around 0.8 V vs Ag/Ag$^+$ in the CVs. The redox wave is based on the redox of Fe(II)/(III). The redox potentials ($E_{1/2}$) of Fe(II)/(III) in **polyFeL1** and **polyFeL2** (or **polyFeL3** and **polyFeL4**) are almost the same. These results indicate that the effect of the spacer on $E_{1/2}$ is quite small. On the other hand, the substituent effect of the ligand on the redox potential is large. The shift to a lower oxidative potential is observed by the introduction of electron-releasing groups (methoxy groups). The shift to higher oxidative potential is confirmed for the introduction of electron-withdrawing groups (bromo groups). These results suggest that the electron-donating groups stabilize the oxidized Fe(III) state and the electron-withdrawing groups destabilize it. Accidentally, we found that the blue polymer film on the working electrode became colorless when an oxidative potential was applied to the film in the CV measurements.

2.3   Blue Electrochromism

Electrochemical color change in a material is called "electrochromism". Since Deb reported an electrochromic (EC) device in 1969, various electrochromic materials (ECMs) including metal oxides, transition metal complexes, and π-conjugated organic molecules and polymers have been investigated. Fe(II)-based metallo-supramolecular polymers are, however, new ECMs.

A thin film of **polyFeL1** is prepared on an indium tin oxide (ITO) electrode and used as a working electrode. When 1.5 V versus Ag/Ag$^+$ is applied to the polymer film in the electrochemical cell shown in Fig. 2.3a, the blue polymer film becomes colorless. This electrochromism is considered to be caused by the change in the oxidation state of the metal ions in the polymer from Fe(II) to Fe(III), because the redox potential of Fe(II)/(III) is 0.77 V versus Ag/Ag$^+$. When 0 V versus Ag/Ag$^+$ is applied to the colorless film, the original blue color reappears. The electrochromism between blue and colorless is reversible. The blue color is the complementary color of the MLCT absorption from the HOMO (the d electron orbital with the highest energy potential in Fe(II) ions) to the LUMO (the π* orbital in the ligand) in the

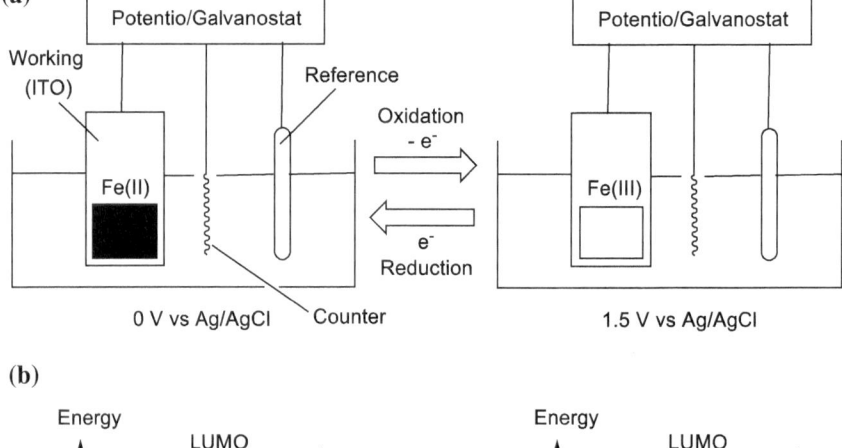

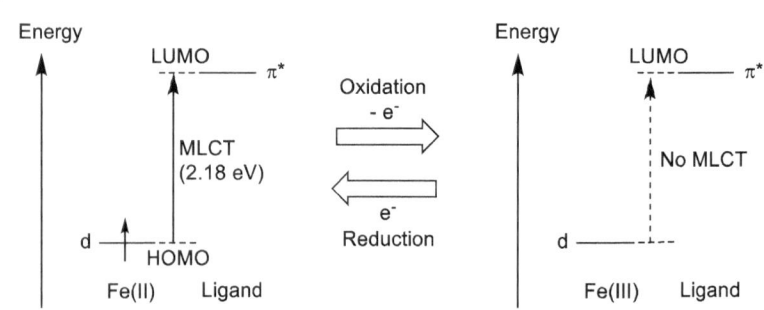

Fig. 2.3 **a** Electrochromism in a thin film of **polyFeL1** on a working electrode and **b** a proposed mechanism of electrochromism in the polymer

complex moieties (Fig. 2.3b). When Fe(II) in the polymer is oxidized to Fe(III) electrochemically, the d electron with the highest energy potential, which contributes to the MLCT absorption, is removed, and the MLCT absorption disappears.

2.4 Electrochromic Devices

Electrochromism in materials has received special attention for display device applications. For example, an electrochromic "smart window" is now built into the newest airplanes (the Boeing 787).

An electrochromic solid-state device incorporating a metallo-supramolecular polymer has been fabricated with a gel electrolyte [5–7]. The typical fabrication procedure is as follows. A thin film of **polyFeL1** is prepared on an ITO-coated glass by spin-coating the polymer in methanol solution (2.0 mg/mL). A gel electrolyte is prepared by mixing poly(methyl methacrylate) (PMMA) (7.0 g), propylene carbonate (PC) (20 mL), and $LiClO_4$ (3.0 g). Another ITO-coated glass is covered with the gel electrolyte. The polymer film and the gel electrolyte layer are stuck together by an insulating film with an arbitrary image, which has been cut off (Fig. 2.4a). When a voltage higher than 2.5 V was applied between the two ITO electrodes in the device [polymer side: positive (+); electrolyte side: negative (−)], a colorless image appeared immediately on the blue polymer film, because the electrochemical oxidation of Fe(II) to Fe(III) occurred in the device (Fig. 2.4b). When the opposite voltage was applied, the colorless part changed to blue owing to the electrochemical reduction of Fe(III) to Fe(II). When the color changes from blue to colorless in the

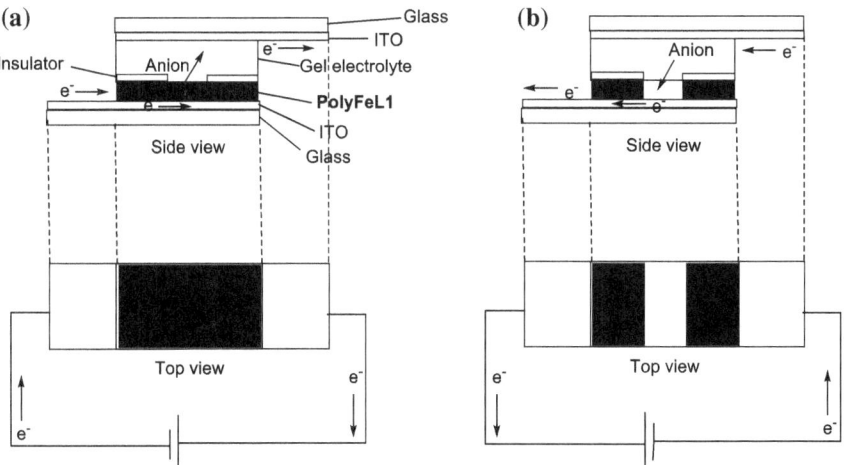

Fig. 2.4 An electrochromic solid-state device using a **polyFeL1** film and the mechanism of color change. **a** Reduction and **b** oxidation states of the polymer in the device

polymer, electrons are removed from the polymer to the electrode. At the same time, anions in the electrolyte layer move to the polymer to neutralize the positive charge generated by the oxidation of Fe(II) to Fe(III). Since the insulating film prevents anion transfer from the electrolyte layer to the polymer film, the part in the polymer film covered with the insulating film does not change color when voltage is applied.

2.5 Electrochromic Properties of Hyperbranched Structures

As described previously, metallo-supramolecular polymers with a linear structure are prepared by the 1:1 complexation of metal ions with ditopic ligands. On the other hand, hyperbranched metallo-supramolecular polymers are synthesized by the further addition of metal ions and tritopic ligands to a solution of the linear metallo-supramolecular polymer (Fig. 2.5) [8]. The degree of branching in the resultant polymer can be controlled by changing the molar ratio of n/m as defined in the figure. When the ratio of m to n is low, a polymer with few branches is formed, and when the ratio of m is high, a highly branched polymer results. Since the metallo-supramolecular polymer backbone is composed of coordinate bonds formed in a reaction that goes to equilibrium, the added tritopic ligands can be introduced to not only the terminal sites of the original, linear polymer, but also the middle part of the polymer chain during the hyperbranching process.

According to the synthetic procedure shown in Fig. 2.5, Fe(II)-based metallo-supramolecular polymers with a hyperbranched structure (**polyFeL1$_{x\%}$L6$_{y\%}$**) are prepared using tris(terpyridine) (**L6**) as a tritopic ligand (Fig. 2.6). The molar ratios of **L1** and **L6** are shown as x% and y%, respectively, in the abbreviation. Molecular weight of the polymers couldn't be determined by GPC measurement because of the decomplexation in the GPC column.

Electrochromism is evaluated on the basis of (1) transmittance difference (ΔT) (=contrast) between the colored state ($T_{colored}$) and the colorless state ($T_{bleached}$), (2) switching times (=speed) for coloring ($t_{coloring}$) and bleaching ($t_{bleaching}$), (3) applied potentials needed to cause the color change, (4) charge quantity needed to cause the color change, (5) coloration efficiency (η), (6) time to maintain the colored or bleached state, and (7) durability for the repeated color changes. A large ΔT is achieved by a low $T_{colored}$ and a high $T_{bleached}$. A low $T_{colored}$ means an absorption with a high intensity, which is characteristic of a material with a high absorption coefficient (ε). The high $T_{bleached}$ suggests the total disappearance of the absorption. Therefore, the total disappearance of an absorption band with high ε by the electrochemical redox of the material results in a large ΔT. As for the switching time (t), the mechanism is more complicated than that for ΔT. The color change in an electrochromic material is caused by the removal/injection of electrons from/to the material to/from an electrode. In electron transfer, ion transfer takes place between the electrochromic material and an electrolyte in the electrochemical cell. Whichever

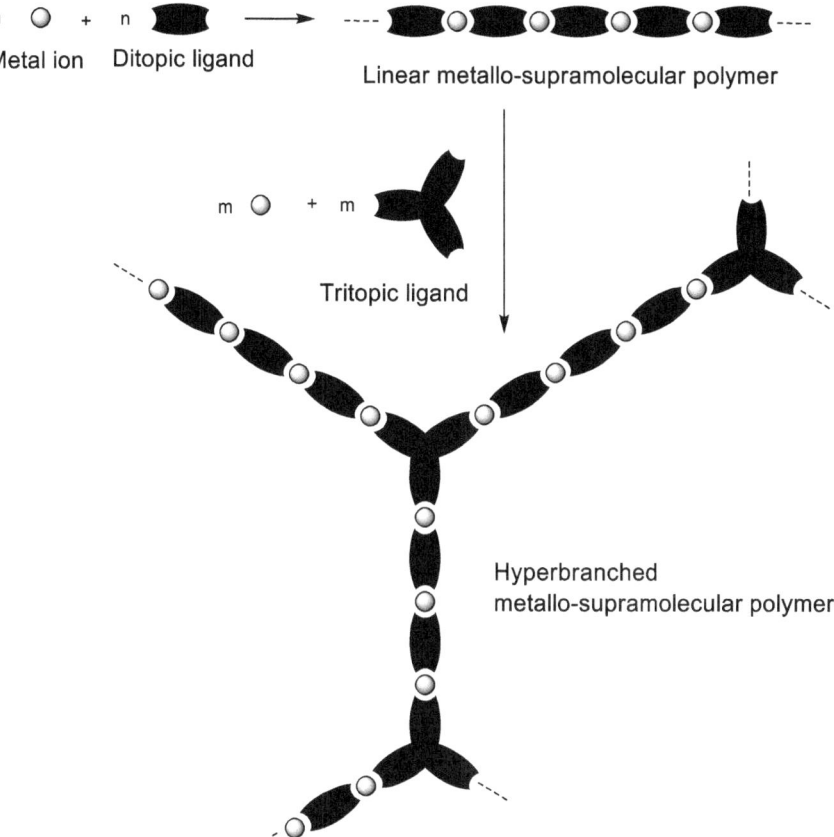

Fig. 2.5 A hyperbranched metallo-supramolecular polymer synthesized by the further addition of metal ions and tritopic ligands to a solution of the linear metallo-supramolecular polymer

is the slower—electron transfer or ion transfer—determines the switching speed for coloring and bleaching. Applied potentials needed for the color changes are based on the redox potential of the moiety, which causes the color change in the material, but higher potentials are often necessary to change the color owing to resistance in the cell. The quantity of electric charge needed for the color change is theoretically the total number of coulombs (C) required to completely oxidize or reduce the redox active sites in the electrochromic material. However, a much larger amount of charge is necessary for electrochromism owing to the internal resistance in the material. The coloration efficiency is an index for evaluating energy consumption during electrochromism. The difference in optical densities (ΔOD) induced as a function of the injected/ejected electronic charge (Q_d) may be measured and is given by Eq. 2.1.

$$\eta = \Delta OD/Q_d = \log(T_{\text{bleached}}/T_{\text{colored}})/Q_d \qquad (2.1)$$

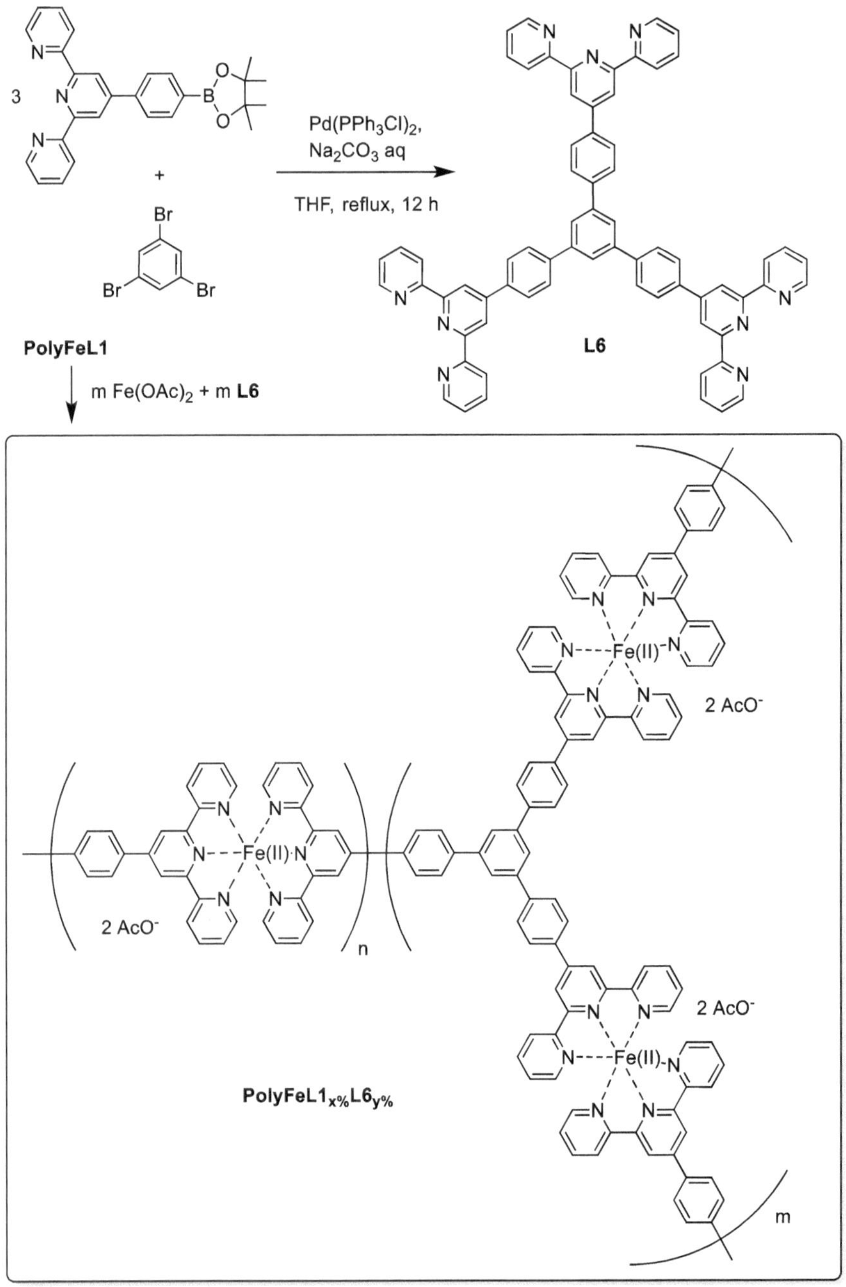

Fig. 2.6 Synthesis of tris(terpyridine) (**L6**) and the polymer (**polyFeL1$_{x\%}$L6$_{y\%}$**)

Here, η (cm^2/C) is the coloration efficiency at a given ΔOD, and $T_{bleached}$ and $T_{colored}$ are the transmittances in the bleached and colored states, respectively. Q_d is calculated from the current change as a function of time while a potential is applied. The memory time for retaining the colored or bleached state is one of characteristic features of electrochromic materials. An electrochromic material with a long memory time is suitable for displays with low energy consumption. High durability with respect to the repeated color changes is an essential property for any application. Preventing the degradation of the material during redox is effective for enhancing the durability.

The Fe(II)-based metallo-supramolecular polymers have relatively excellent electrochromic properties. In addition, electrochromic properties are affected by morphology, because morphological differences change the electron/ion transfer behavior that produces the electrochromism. Electron transfer between an electrode and an electrochromic layer is generally faster than ion transfer between an electrochromic layer and a gel electrolyte layer in a cell. Therefore, it is expected that smooth ion transfer accelerates the switching speed. One way to smoothen the ion transfer is by increasing the surface area in the electrochromic layer in contact with the electrolyte. In the case of linear polymers, the formation of bundle structures due to the stacking of the polymer chains is anticipated. On the other hand, a film of hyperbranched polymers is expected to show a porous structure, because less interaction between the polymer chains prevents the chains from stacking.

Atomic force microscopy (AFM) is a useful method for investigating surface morphology, especially of electrically insulating materials. The AFM images of the film surface in Fe(II)-based metallo-supramolecular polymers bearing a hyperbranched structure (**polyFeL1$_{x\%}$L6$_{y\%}$**) show highly porous structures (pore size: approximately 30–50 nm in diameter) as the ratio of **L6** increases from 0 to 15%. The porousness becomes worse as the ratio is increased to more than 18%. The most highly porous structure appears in the polymer in which the molar ratio of **L6** is 15% (**polyFeL1$_{85\%}$L6$_{15\%}$**). Interestingly, the most porous polymer film exhibits the best electrochromic performance (coloring time: 0.19 s; bleaching time: 0.36 s; transmittance change (ΔT): 50.7%; coloration efficiency (η): 383.4 cm^2/C) (Table 2.2). The electrochemical redox of Fe(II)/(III) in the polymer film is diffusion-controlled, as supported by the linear relationship between the current and the square root of the scan rate in cyclic voltammetry. The porous structure contributes to the smooth transfer of ions during redox and demonstrates excellent electrochromic properties.

2.6 Ionic Conductivity of Hyperbranched Structures

As described in the previous section, the introduction of a tritopic ligand changes the polymer structure from linear to hyperbranched. As the ratio of a tritopic ligand to the total amount of ditopic and tritopic ligands increases, preparing the polymer film by spin coating becomes difficult because of the decrease in the solubility. Therefore, 20% is the maximum ratio of the tritopic ligand to the total amount of

Table 2.2 Electrochromic properties of Fe(II)-based metallo-supramolecular polymers with linear or hyperbranched structures

	$T_{bleached}$ (%)	$T_{colored}$ (%)	ΔT (%)	$t_{coloring}$ (s)	$t_{bleaching}$ (s)	Charge/Discharge (mC)	η (cm^2/C)
PolyFeL1$_{100\%}$	93.2	51.6	41.6	0.31	0.58	1.46/1.44	263.8
PolyFeL1$_{95\%}$L6$_{5\%}$	92.5	43.9	48.6	0.21	0.51	1.69/1.65	287.2
PolyFeL1$_{90\%}$L6$_{10\%}$	94.0	46.5	47.5	0.23	0.52	1.38/1.35	332.2
PolyFeL1$_{85\%}$L6$_{15\%}$	91.6	40.9	50.7	0.19	0.36	1.37/1.34	383.4
PolyFeL1$_{82\%}$L6$_{18\%}$	78.8	54.8	24.0	0.37	0.60	1.69/1.65	140.0
PolyFeL1$_{80\%}$L6$_{20\%}$	81.6	54.4	27.2	0.37	0.62	2.02/1.97	130.7

The transmittances of the MLCT absorption (λ_{max}) in the bleached ($T_{bleached}$) and colored states ($T_{colored}$) of the polymer film coated on an ITO glass are measured by in situ UV-vis spectroscopy at 0 or 1.2 V versus Ag/Ag$^+$ with an interval time of 5 s (electrolyte: 0.1 M LiClO$_4$/acetonitrile; ITO working area: 1 × 1.5 cm). The transmittance difference (ΔT) is calculated from $T_{bleached}$ and $T_{colored}$. The times for coloring and bleaching ($t_{coloring}$ and $t_{bleaching}$) are defined as the time taken for ΔT to change by 95%. The charge/discharge values are calculated from the integration of the coulomb number in the current response during the redox. The coloration efficiency (η) is defined as the relationship between the charge used and the change in ΔT

ligand to maintain sufficient solubility for a uniform film preparation of the polymer. However, if high solubility is not required for the applications, the simple mixing of metal ions and tritopic ligands is another way of obtaining a hyperbranched metallo-supramolecular polymer (Fig. 2.7) [9]. To obtain a fully hyperbranched structure, the molar ratio of metal ion to tritopic ligand should be 3:2.

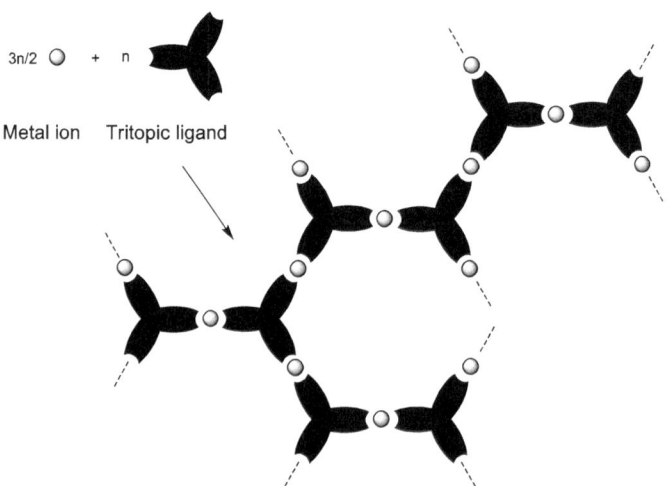

Hyperbranched metallo-supramolecular polymer

Fig. 2.7 Synthesis of hyperbranched metallo-supramolecular polymer by the complexation of metal ions with tritopic ligands

The hyperbranched polymers **polyFeL6**, **polyCoL6**, and **polyNiL6** are prepared by the complexation of tris(terpyridine) (**L6**) with Fe(II), Co(II), and Ni(II), respectively (Fig. 2.8), as follows. A 2:3 ratio molar of **L6** and M(OAc)$_2$ (M: Fe, Co, Ni)

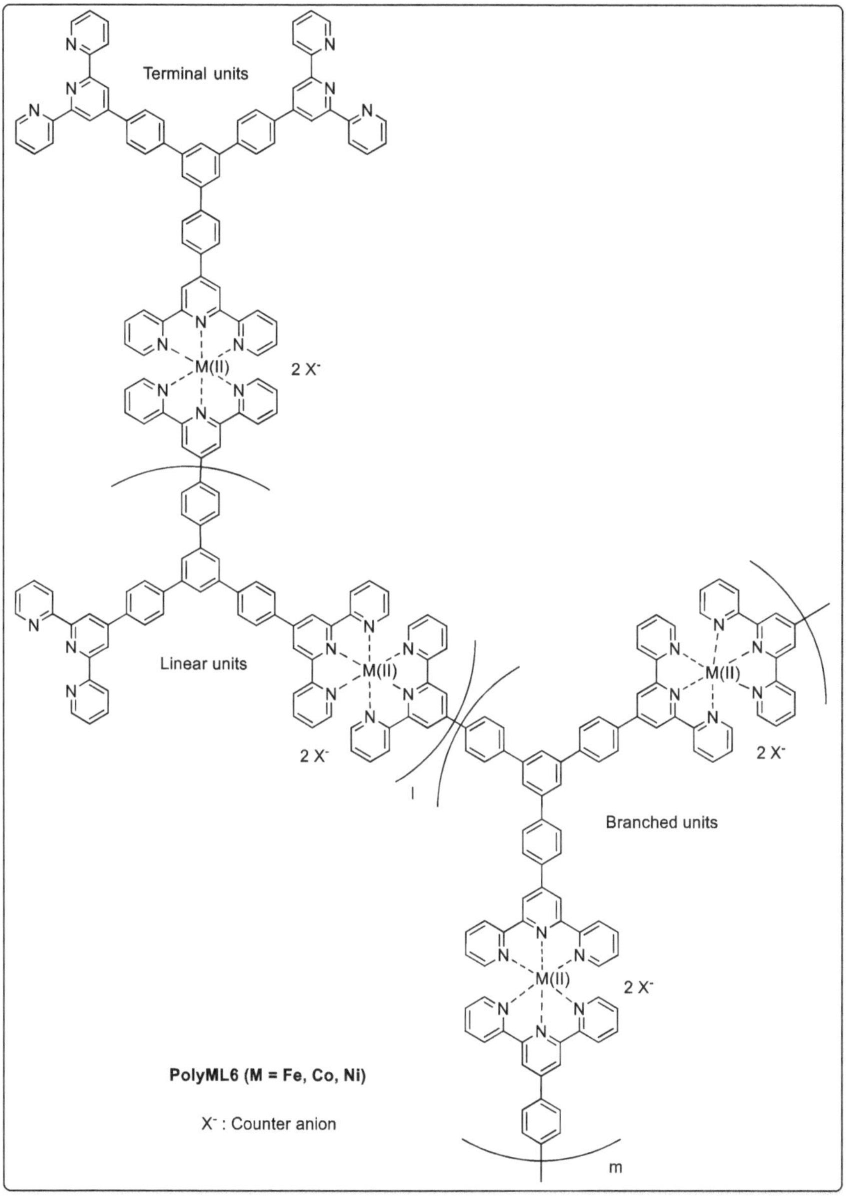

Fig. 2.8 Hyperbranched metallo-supramolecular polymers composed of metal ions (Fe(II), Co(II), or (Ni)) and tris(terpyridine) (**L6**) (**polyFeL6**, **polyCoL6**, and **polyNiL6**)

is refluxed in argon-saturated acetic acid (ca. 10 mL of solvent per mg of **L6**) for 24 h. The reaction solution is cooled to room temperature and the precipitate is collected, washed with chloroform, water, and diethyl ether, and dried under vacuum overnight to yield the corresponding polymers quantitatively. The polymer structure is classified by branched, linear, and terminal units as shown in the figure.

Compared with the hyperbranched polymers, the corresponding linear polymers (**polyFeL1**, **polyCoL1**, and **polyNiL1**) are synthesized by the 1:1 complexation of **L1** and the metal ions. The XRD powder spectrum of **polyFeL6** shows broader peaks than that of **polyFeL1**. It indicates that the hyperbranched polymer is more amorphous than the linear polymer, probably because the stacking of the polymer chains is prevented in the hyperbranched polymer.

Ionic conduction has received much attention in terms of both scientific interest and industrial applications such as electrolytes in fuel cells. We observed ionic conduction in metallo-supramolecular polymers quite by accident. Since the details are introduced in Chap. 8, this section focuses on only the effect of polymer structure on ionic conduction. The dc I–V measurements on the polymers showed only negligible current response, which suggested quite a low electrical conductivity of the polymers. Then, ac impedance analysis was carried out at various relative humidity levels. A pellet, prepared by pressing the polymer powder, was used for the measurement, because a polymer pellet gives more reliable data than a polymer film. The polymer pellet was held between the two electrodes of a sample holder. The whole electrode setup was then placed inside a humidity- and temperature-controlled chamber. The complex impedance plots (Nyquist plots) were obtained at room temperature and 98 %RH for the polymers from the impedance measurements (20 MHz to 1 Hz). The ionic conductivity was obtained from the x-axis intercept of the plot at high frequencies. The ionic conductivities of the hyperbranched polymers were much higher than those of the corresponding linear polymers (Table 2.3). These results indicate that the ionic conductivity is affected more by structure than by the metal ion species. Hydrophilicity of metal complex moieties in the polymer gathers water molecules

Table 2.3 Ionic conductivity and activation energy of Fe(II)-, Co(II)-, and Ni(II)-based metallo-supramolecular polymers with a linear or hyperbranched structure

	Polymer structure	Ion conductivity at 98%RH at r.t. (mS/cm)	Activation energy (E_a) (eV)
PolyFeL1	Linear	0.32	2.5
PolyFeL6	Hyperbranched	5.7	1.56
PolyCoL1	Linear	5.0×10^{-2}	3.0
PolyCoL6	Hyperbranched	4.8	1.35
PolyNiL1	Linear	2.0×10^{-2}	3.3
PolyNiL6	Hyperbranched	1.4	1.3

The activation energy (E_a) is determined by measuring the ionic conductivity at different temperatures ranging from 20 to 60 °C at 98 %RH. The E_a is calculated from the slope of the Arrhenius plot

under humid conditions and forms water channels along the polymer chains. It is considered that the water channels expand from one-dimension (1D) to 3D by changing the linear polymer structure to the hyperbranched one. Multiple-pathways of ion conduction caused by the 3D water channels enhance the ionic conductivity of the hyperbranched polymers. A smoother ion transfer in the hyperbranched polymers than in the linear ones is also supported by lower activation energies.

2.7 Electrochemical Emission Control of Anionic Dye

Counter anions (acetate, $^{-}$OAc) in **polyFeL1** can be exchanged for fluorescent sulforhodamine B (SRB) anions by the addition of an aqueous solution (10 mL) of SRB sodium salt (17.70 mg/mL) to an aqueous solution (80 mL) of **polyFeL1** (0.27 mg/mL) (Fig. 2.9a) [10]. The precipitated product is filtered, washed with pure water, chloroform, and diethylether, and then dried *in vacuo*. The polymer with SRB (**polyFeL1-SRB**) was obtained in an 88.8% yield. An electrochemical emission switching device containing **polyFeL1** and **polyFeL1-SRB** was fabricated using a gel electrolyte and two ITO electrodes (Fig. 2.9b), as follows. A polymer solution is prepared by mixing methanol solutions of **polyFeL1** (1.1 mg/mL) and **polyFeL1-SRB** (0.1 mg/mL). A film including the two polymers is prepared by spin-coating 400

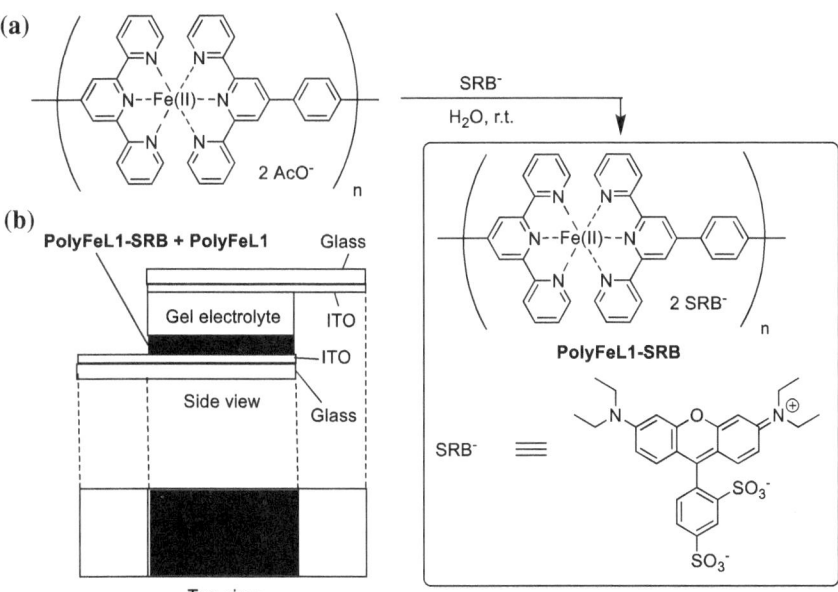

Fig. 2.9 a An exchange reaction of the counter anions in **polyFeL1**. **b** An electrochemical emission switching device including a film of **polyFeL1** and **polyFeL1-SRB**

μL of the mixed polymer solution on an ITO glass (25 × 75 mm). A gel electrolyte is prepared by mixing propylene carbonate (PC) (20 mL), poly(methyl methacrylate) (PMMA) (20 g), and $LiClO_4$ (1.6 g), followed by drying on a glass slide. The device shows electrochromism based on the appearance/disappearance of the MLCT absorption triggered by the redox of Fe(II)/(III) by applying −2.8 and 2.8 V, respectively. Interestingly, the emission of SRB in the device under UV light irradiation (365 nm) appears at 584 nm when −2.8 V is applied, and it is quenched when 2.8 V is applied. This emitting/quenching behavior is reversible. The quenching of emission occurs because the MLCT band absorbs the excitation light at 560 nm and the emission from the SRB.

References

1. Higuchi M, Kurth DG (2007) Electrochemical functions of metallo-supramolecular nano-materials. Chem Rec 7:203–209. https://doi.org/10.1002/tcr.20118
2. Han FS, Higuchi M, Kurth DG (2007) Diverse synthesis of novel bisterpyridines via suzuki-type cross-coupling. Org Lett 9:559–562. https://doi.org/10.1021/ol062788h
3. Han FS, Higuchi M, Kurth DG (2007) Metallo-supramolecular polymers based on functionalized bis-terpyridines as novel electrochromic materials. Adv Mater 19:3928–3931. https://doi.org/10.1002/adma.200700931
4. Han FS, Higuchi M, Kurth DG (2008) Metallo-supramolecular polyelectrolytes self-assembled from various pyridine ring substituted bis-terpyridines and metal ions: photophysical, electrochemical and electrochromic properties. J Am Chem Soc 130:2073–2081. https://doi.org/10.1021/ja710380a
5. Higuchi M, Akasaka Y, Ikeda T, Hayashi A, Kurth DG (2009) Electrochromic solid-state devices using organic-metallic hybrid polymers. J Inorg Organomet Polym Mater 19:74–78. https://doi.org/10.1007/s10904-008-9243-7
6. Higuchi M (2009) Electrochromic organic–metallic hybrid polymers: fundamentals and device applications. Polym J 41:511–520. https://doi.org/10.1295/polymj.PJ2009053
7. Higuchi M (2014) Stimuli-responsive metallo-supramolecular polymer films: design, synthesis and device fabrication. J Mater Chem C 2:9331–9341. https://doi.org/10.1039/c4tc00689e
8. Hu CH, Sato T, Zhang J, Moriyama S, Higuchi M (2014) Three-dimensional Fe(II)-based metallo-supramolecular polymers with electrochromic properties of quick switching, large contrast, and high coloration efficiency. ACS Appl Mater Interfaces 6:9118–9125. https://doi.org/10.1021/am5010859
9. Pandey RK, Hossain MD, Sato T, Rana U, Moriyama S, Higuchi M (2015) Effect of a three-dimensional hyperbranched structure on the ionic conduction of metallo-supramolecular polymers. RSC Adv 5:49224–49230. https://doi.org/10.1039/c5ra07217d
10. Suzuki T, Sato T, Zhang J, Kanao M, Higuchi M, Maki H (2016) Electrochemically switchable photoluminescence of an anionic dye in a cationic metallo-supramolecular polymer. J Mater Chem C 4:1594–1598. https://doi.org/10.1039/c5tc03071d

Chapter 3
Ru(II)-Based Metallo-Supramolecular Polymers

3.1 Synthesis

Bis(terpyridine)s are widely used as ditopic ligands for metallo-supramolecular polymer synthesis because of their high coordination ability. In particular, transition metal
ions with an octahedral coordination such as Fe(II) and Ru(II) fit into the tridentate
coordination sites on the ligands. Similar to the Fe(II)-based polymers, Ru(II)-based
metallo-supramolecular polymers (**polyRuL1-5**) are prepared by the 1:1 complexation of a Ru(II) salt and bis(terpyridine)s (**L1-5**) (Fig. 3.1) [1]. The complexation
conditions are different from those for the Fe(II)-based polymers. The typical synthetic procedure is as follows. Equimolar amounts of **L1** and $RuCl_2(DMSO)_4$ are
stirred at 130 °C in argon-saturated absolute ethylene glycol (*ca.* 10 mL of solvent
per mg of **L1**) for 24 h. After the solution is cooled to room temperature, THF is
added until the solution becomes colorless. The precipitated polymer is collected by
filtration, washed twice with THF, and then dried under vacuum overnight to give
polyRuL1 (>95%). The complexation in the polymer synthesis is confirmed by the
color change of the reaction mixture to red, because the complexes have MLCT
absorptions.

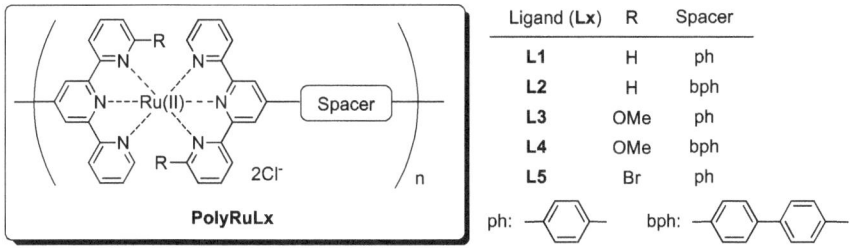

Fig. 3.1 a Ru(II)-based metallo-supramolecular polymers (**polyRuL1-5**)

© National Institute for Materials Science, Japan 2019   29
M. Higuchi, *Metallo-Supramolecular Polymers*, NIMS Monographs,
https://doi.org/10.1007/978-4-431-56891-9_3

In general, the chain length of metallo-supramolecular polymers in solution is difficult to determine because the position of equilibrium involving the coordination of the polymer changes with the polymer chain length. However, the linear polymer structure of **polyRuL1** can be observed as bright threads in an atomic force microscopy (AFM) image of solid-state **polyRuL1** cast on a Si substrate from the dilute methanol solution. The observed height (1–2 nm) indicates that the bright threads are single polymer chains or small bundles of several polymer chains, because the polymer chain width estimated by molecular modeling is about 1 nm. In the AFM image, the polymer chain length is more than 500 nm. Since the distance between two Ru ions in the polymer is estimated to be about 1.6 nm by molecular modeling, the calculated degree of polymerization (DP) should be more than 300, which corresponds to a molecular weight of 2.1×10^5 Da.

3.2 Red Electrochromism

Similar to **polyFeL1-5**, the optical properties of **polyRuL1-5** are related to the MLCT absorption. In the UV-vis spectra of **polyRuL1-5**, the MLCT absorption of the Ru(II) complex moieties appears around 520 nm. The maximum wavelength (λ_{max}) and absorption coefficient (ε) are summarized in Table 3.1. Similar to the Fe(II) polymers, the introduction of a biphenyl group into the ligand as a spacer leads to a blue shift of the absorption. The introduction of methoxy groups into the ligand causes a large decrease in ε and a red shift of λ_{max} by 20 nm. The introduction of bromo groups results in a large decrease in ε and a blue shift of λ_{max}. The different substituent effects on λ_{max} and ε between **polyFeL1-5** and **polyRuL1-5** depend on subtle differences in the structures of the metal complexes. As an overall tendency, the MLCT absorption in **polyRuL1-5** appears at a shorter wavelength and has a higher ε than that in

Table 3.1 Optical and electrochemical properties of Ru(II)-based metallo-supramolecular polymers (**polyRuL1-5**)

	Maximum wavelength (λ_{max})[a] (nm)	Absorption coefficient (ε)[a], $\times 10^4$	Redox potential ($E_{1/2}$)[b], V versus Ag/Ag$^+$
PolyRuL1	513	4.10	0.95
PolyRuL1 in MeOH	508	4.00	–
PolyRuL2	502	3.55	0.95
PolyRuL3	536	2.54	0.84
PolyRuL4	524	2.25	0.85
PolyRuL5	507	3.25	1.16

[a]Solvent: methanol–water (4:1). [b]Working electrode: glassy carbon; counter electrode: Pt wire; reference electrode: Ag/Ag$^+$; electrolyte: 0.10 M n-Bu$_4$NClO$_4$/acetonitrile; scan rate: 100 mV/s

polyFeL1-5, probably because of the stronger π-backbonding of **L1-5** to Ru(II) than to Fe(II) and a stronger dynamic chelate effect of **L1-5** to Ru(II) than to Fe(II).

In cyclic voltammograms (CVs) of **polyRuL1-5**, a redox wave is observed owing to the redox of Ru(II)/(III). The results clearly show the shift to a lower oxidative potential by the introduction of electron-releasing groups (methoxy groups) and a shift to a higher oxidative potential by the introduction of electron-withdrawing groups (bromo groups), because the strong electron-donating groups stabilize the oxidized Ru(III) state and the electron-withdrawing groups destabilize it. The trend is quite similar to that observed in **polyFeL1-5**.

The film of **polyRuL1-5** on an ITO glass shows electrochromic properties from red to colorless when an oxidative potential is applied to the film in an electrolyte solution, because Ru(II) is oxidized to Ru(III) electrochemically and the MLCT absorption disappears. The colorless film shows the original red color when a reductive potential is applied to the film.

3.3 Fe(II)/Ru(II)-Based Heterometallo-Supramolecular Polymers

Heterometallo-Supramolecular polymers are metallo-supramolecular polymers containing more than two metal ion species. The introduction of more than two metal ion species to a metallo-supramolecular backbone itself is not difficult. The polymers may be obtained by mixing different metal salts and a ditopic ligand in solution. However, precise control of the metal sequence is quite difficult, because simple mixing gives a polymer with the metal ions randomly introduced. To control the position of the two metal ion species in the polymer, we focused on the different reaction conditions for metal ions with the ligand. If the reaction conditions for the ligand are significantly different between the two metal ion species, the polymer with two metal ion species introduced alternately can be prepared by two-step complexations of the metal ions with the ligand (Fig. 3.2a) [2].

At first, the 1:2 complex of one metal with a ligand is prepared under the complexation conditions; then the polymer with two metal ion species introduced alternately is obtained by the complexation of the 1:2 complex with another metal ion species. In the two-step complexations, the complexation with the more severe reaction conditions should be the first complexation. If the complexation with the milder reaction conditions is selected as the first complexation, exchange between the two metal ion species can occur during the second complexation with the more severe reaction conditions. According to this synthetic strategy, a series of Fe/Ru-based heterometallo-supramolecular polymers are synthesized by the stepwise coordination of Fe(II) and Ru(II) ions to **L1**. The molar ratio of Fe(II) to Ru(II) in the polymer can be controlled by changing the molar ratio in the polymerization. The polymer with the 1:1 molar ratio of Fe(II) to Ru(II) (**polyFe$_{0.5}$Ru$_{0.5}$L1**) is shown in Fig. 3.2b.

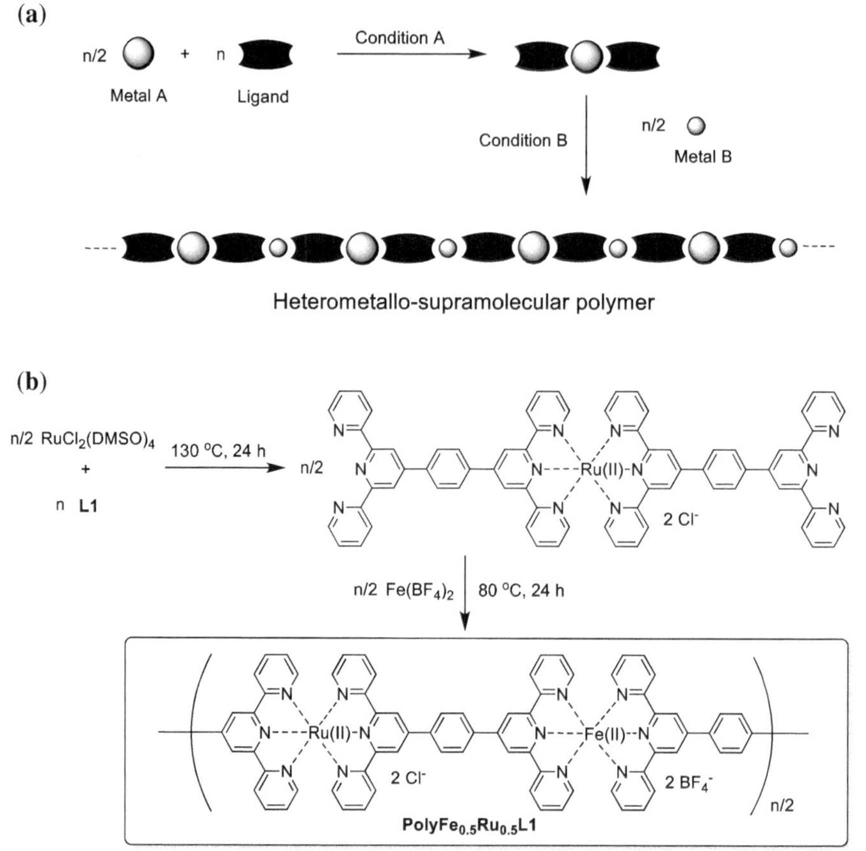

(a)

Heterometallo-supramolecular polymer

(b)

Fig. 3.2 a Synthetic strategy for heterometallo-supramolecular polymers. **b** Synthesis of Fe(II)/Ru(II)-based heterometallo-supramolecular polymers (**polyFe$_{0.5}$Ru$_{0.5}$L1**)

The molar ratio of Fe(II) to Ru(II) in the heterometallo-supramolecular polymer can be changed by changing the seeding molar ratio of the metal salts. The polymers with the 3:1, 1:1, and 1:3 molar ratios of Fe(II) to Ru(II) (**polyFe$_{0.75}$Ru$_{0.25}$L1**, **polyFe$_{0.5}$Ru$_{0.5}$L1**, and **polyFe$_{0.25}$Ru$_{0.75}$L1**, respectively) are synthesized by seeding the calculated molar ratios of Fe(BF$_4$)$_2$ to RuCl$_2$(DMSO)$_4$ (Table 3.2). The preparation is as follows. First, **L1** and RuCl$_2$(DMSO)$_4$ are added in argon-saturated absolute ethylene glycol (EG) and stirred at 130 °C for 24 h. After the reaction, Fe(BF$_4$)$_2$ dissolved in EG is added into the reaction mixture and stirred at 80 °C for 24 h. Following the same purification method used for **polyRuL1**, the polymers are obtained in >90% yields. Molecular weight of the polymers couldn't be determined by GPC measurement because of the decomplexation in the GPC column.

The redox potentials ($E_{1/2}$) of Fe(II)/(III) and Ru(II)/(III) in the polymers are determined by cyclic voltammetry (CV) and summarized in Table 3.3. The heterometallo-supramolecular polymers have two redox waves based on the redox of Fe(II)/(III)

Table 3.2 Seeding molar ratios of Fe(II), Ru(II), and **L1** in the synthesis of Fe(II)/Ru(II)-based heterometallo-supramolecular polymers

	Fe(BF$_4$)$_2$ (mmol)	RuCl$_2$(DMSO)$_4$ (mmol)	**L1** (mmol)
PolyFeL1	0.2	0	0.2
PolyFe$_{0.75}$Ru$_{0.25}$L1	0.15	0.05	0.2
PolyFe$_{0.5}$Ru$_{0.5}$L1	0.1	0.1	0.2
PolyFe$_{0.25}$Ru$_{0.75}$L1	0.05	0.15	0.2
PolyRuL1	0	0.2	0.2

Table 3.3 Oxidation and reduction potentials of Fe(II)/Ru(II)-based heterometallo-supramolecular polymers

	E_{ox} of Fe(II)/(III) (mV)	E_{red} of Fe(II)/(III) (mV)	$E_{1/2}$ of Fe(II)/(III) (mV)	E_{ox} of Ru(II)/(III) (mV)	E_{red} of Ru(II)/(III) (mV)	$E_{1/2}$ of Ru(II)/(III) (mV)
PolyFeL1	768	741	755	–	–	–
PolyFe$_{0.75}$Ru$_{0.25}$L1	773	744	759	922	903	913
PolyFe$_{0.5}$Ru$_{0.5}$L1	781	746	764	933	904	919
PolyFe$_{0.25}$Ru$_{0.75}$L1	785	753	769	945	909	927
PolyRuL1	–	–	–	949	916	933

Working electrode: ITO glass (active area: 1×1 cm^2); reference electrode: Ag/Ag$^+$; counter electrode: Pt wire; electrolyte: 0.1 M LiClO$_4$; scan rate: 20 mV/s; polymer film preparation on the ITO glass: spray coating

and Ru(II)/(III). The data show that the redox potential of Fe(II)/(III) is positively shifted as the ratio of Ru ions in the polymers increases, and that of Ru(II)/(III) is negatively shifted as the ratio of Fe ions in the polymers increases. These shifts clearly suggest that the intramolecular metal–metal interactions occur between the adjacent Fe and Ru ions through the π-conjugated spacer in the ligand.

3.4 Multi-Color Electrochromism

The colors of **polyFeL1** and **polyRuL1** in solution are blue and red, because the complementary color of the MLCT absorption at 585 or 513 nm in the Fe(II)- or Ru(II)-terpyridine complex moieties is seen, respectively. Since the Fe(II)/Ru(II)-based heterometallo-supramolecular polymers (**polyFe$_{0.75}$Ru$_{0.25}$L1**, **polyFe$_{0.5}$Ru$_{0.5}$L1**, and **polyFe$_{0.25}$Ru$_{0.75}$L1**) have both absorptions, the polymers display different colors (bluish purple, purple, and reddish purple, respectively) based on the different absorbance ratios of the two MLCT absorptions. Interestingly, the heterometallo-supramolecular polymers show multi-color electrochromism by changing the applied potential from 0 to 1.1 V versus Ag/Ag$^+$. The multi-color electrochromism in a **polyFe$_{0.5}$Ru$_{0.5}$L1** film (purple at 0 V vs. Ag/Ag$^+$, orange at 0.9 V, and light green at 1.2 V) was investigated by in situ UV–vis spectral measurements, which monitored

Table 3.4 Electrochromic properties of a **polyFe$_{0.5}$Ru$_{0.5}$L1** film when switched between 0 and 0.9 V, and 0 and 1.1 V at 5 s intervals

Applied potential, V versus Ag/Ag$^+$	λ_{max} (nm)	$t_{coloring}$ (s)	$t_{bleaching}$ (s)	ΔT (%)	Q_d (mC/cm$^{2)}$	η (cm^2/C)
0–0.9	585	0.4	1.5	37.93	1.60/2.19	188.2
0–1.1	508	0.4	1.5	68.70	2.66/2.72	242.1

The transmittance changes of the MLCT absorption at the maximum wavelength (λ_{max}) in the polymer are monitored while a potential is applied. The transmittance difference (ΔT) is calculated from the transmittances in the colored and bleached states ($T_{colored}$ and $T_{bleached}$). The charge/discharge amount (Q_d) is calculated from the time–current curve in chronoamperometry. The coloration efficiency (η) is calculated from $T_{colored}$, $T_{bleached}$, and Q_d using Eq. 2.1. Working electrode: ITO glass (active area: 1×1 cm^2); reference electrode: Ag/Ag$^+$; counter electrode: Pt wire; electrolyte: 0.1 M LiClO$_4$/acetonitrile; bleaching and coloring times, the time needed for 95% of ΔT to change. The polymer films are prepared on the ITO glass by spray coating the polymer in methanol solution

the absorbance changes while applying a potential. At 0.7 V versus Ag/Ag$^+$, only the MLCT absorption at 585 nm attributed to the Fe(II) complex moieties decreased slightly. At 0.8 V versus Ag/Ag$^+$, the absorption decreased greatly, and that of Ru(II) at 508 nm also began to decrease. At 0.9 V versus Ag/Ag$^+$, the absorption of Fe(II) almost disappears, but that of Ru(II) still remains. Finally, the absorption of Ru(II) totally disappears at 1.1 V versus Ag/Ag$^+$. Since the oxidation potentials of Fe(II) and Ru(II) ions in the polymer are different, the two MLCT absorptions disappear stepwise due to the electrochemical oxidation of Fe(II) to Fe(III) and the subsequent oxidation of Ru(II) to Ru(III). This electrochromic performance is summarized in Table 3.4. The polymer films show high optical contrast and very fast response times: the transmittance changes (ΔT) at 508 and 585 nm are 68% and 37%, respectively, and the color changes end at 0.4 and 1.5 s for coloring and bleaching, respectively. In addition, high durability for repeated color changes is confirmed by applying 0 and 0.9 V repeatedly with only 0.7 and 1.8% charge loss in 5,000 and 10,000 cycles, respectively.

3.5 Flexible Electrochromic Devices

Fe(II)- and Ru(II)-based metallo-supramolecular polymers are soluble in polar solvents such as water and methanol. The polymer film can be prepared using spin coating and spray coating on an ITO electrode. Inkjet printing is another method of preparing the polymer film. Mixed color films of blue and red are prepared on ITO substrates (ITO glass: 6.8 Ω/□; flexible ITO-PEN (polyethylene-naphthalate): 35 Ω/□) by inkjet printing different ratios of **polyFeL1** and **polyRuL1** (4/0, 3/1, 2/2, 1/3, 0/4) (Fig. 3.3a) [3]. A 1.0 wt% methanol solution of each polymer is prepared and diluted with an equal volume of deionized water to reduce nozzle clogging problems in the inkjet-printing process. The inks are placed in two cartridges and printed one after another in the printing process for the color-mixing thin films. Droplets 70 μm in diameter are ejected from a nozzle 50 μm in diameter at a speed of 1.01 m/s and a frequency of 500 Hz with a dot spacing of 50 μm for all printed patterns. The substrate is heated at 35 °C during printing. The devices are fabricated with a transparent solid-state electrolyte thin film (Fig. 3.3b). The electrolyte solution is prepared by mixing poly(vinylidenefluoride-co-hexafluoropropylene) (PVDF-HFP), 1-ethyl-3-methylimidazolium-bis(trifluoromethylsulfonyl) amide (EMIBTI), and acetone (weight ratio: 1:4:7). The solution is poured into a glass Petri dish and dried overnight in a vacuum oven at 70 °C for 24 h. The dried thin film (500 μm thick) is then peeled off and cut to the proper size for the device fabrication.

Cyclic voltammograms of the inkjet-printed polymer films show only small changes in the redox potentials of Fe(II)/(III) and Ru(II)/(III) when the molar ratio of **polyFeL1** and **polyRuL1** is changed in the printing (Table 3.5), unlike the case of Fe/Ru-based heterometallo-supramolecular polymers, in which Fe and Ru ions electronically interact through the π-conjugated ditopic ligand. It is considered that most of the two polymers exist individually in the polymer film prepared by the inkjet-printing. However, the slight shifts of redox potential shown in Table 3.5 indicate that a small amount of block-copolymer with both Fe and Ru ions was formed due to the equilibrium reaction in solution during the inkjet-printing. The solid-state devices show multicolored electrochromic behavior. The effect of bending the electrochromic devices is investigated using flexible solid-state devices with **polyFeL1**

(a)

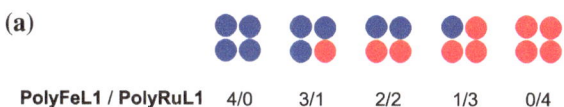

PolyFeL1 / PolyRuL1   4/0   3/1   2/2   1/3   0/4

(b)

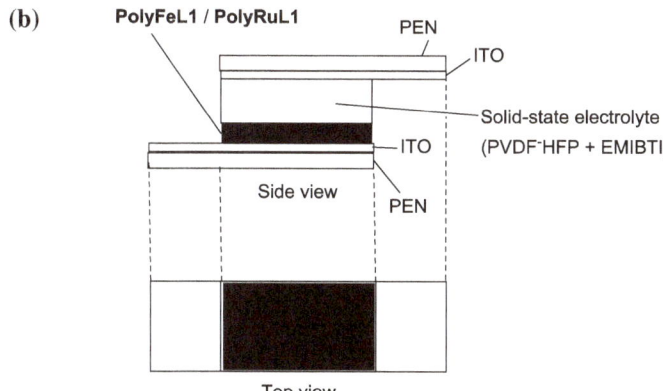

Fig. 3.3   a Dot images of inkjet printing with **polyFeL1** and **polyRuL1** at different ratios from 4/0 to 0/4. **b** A flexible solid-state device structure with a printed electrochromic layer

Table 3.5 Oxidation and reduction potentials of **polyFeL1** and **polyRuL1** in inkjet-printed polymer films

PolyFeL1/PolyRuL1	E_{ox} of Fe(II)/(III) (mV)	E_{red} of Fe(II)/(III) (mV)	$E_{1/2}$ of Fe(II)/(III) (mV)	E_{ox} of Ru(II)/(III) (mV)	E_{red} of Ru(II)/(III) (mV)	$E_{1/2}$ of Ru(II)/(III) (mV)
4/0	771	718	745	–	–	–
3/1	777	718	748	931	899	915
2/2	778	719	749	936	898	917
1/3	780	719	750	940	895	918
0/4	–	–	–	945	896	921

Working electrode: ITO glass (active area: 1×1 cm^2); reference electrode: Ag/Ag$^+$; counter electrode: a Pt sheet (1 cm in width and 2 cm in length); electrolyte: 0.1 M LiClO$_4$, scan rate: 20 mV/s; polymer film preparation on the ITO glass: inkjet printing

Table 3.6 Electrochromic properties of flexible solid-state devices with **polyFeL1** or **polyRuL1**

Polymer	Device state	λ_{max} (nm)	$t_{coloring}$ (s)	$t_{bleaching}$ (s)	ΔT (%)	Q_d (mC/cm^2)	η (cm^2/C)
PolyFeL1	Flat	580	2.0	26	40.1	2.67/2.86	445
PolyFeL1	Bent	580	2.0	21	30.1	2.88/2.98	381
PolyRuL1	Flat	515	0.5	27	32.3	2.10/2.77	521
PolyRuL1	Bent	515	2.0	21	29.9	2.46/2.83	439

The transmittance change of the MLCT absorption at the maximum wavelength (λ_{max}) in the polymer is monitored while a voltage is applied. The voltages are switched between -3.0 V and 3.0 V at 50 s intervals. The transmittance difference (ΔT) is calculated from the transmittances in the colored and bleached states ($T_{colored}$ and $T_{bleached}$). The charge/discharge amount (Q_d) is calculated from the time-current curve in chronoamperometry. The coloration efficiency (η) is calculated from $T_{colored}$, $T_{bleached}$, and Q_d using Eq. 2.1. Electrode: ITO-PEN (active area: 1×1 cm^2); solid-state electrolyte: PVDF-HFP + EMIBTI; bleaching and coloring times, the time needed for 95% of ΔT to change. The polymer films are prepared on the ITO-PEN by inkjet printing

or **polyRuL1** (Table 3.6). The use of PEN with higher resistance than ITO as the electrode results in longer response times ($t_{coloring}$ and $t_{bleaching}$). When the properties of the devices are compared between the flat and bent states, the performance in the bent state is worse than that in the flat state, but the properties in the bent state are still good.

3.6 Luminescence

PolyRuL1-5 have the ligand-centered (LC) and MLCT absorptions (Table 3.7). Luminescence appears owing to the excitation of the LC and MLCT bands, but the quantum yield (Φ_{lum}) is low ($<10^{-6}$) at room temperature (Table 3.8). At 77 K, however, a strong luminescence from the excited MLCT absorption is observed [4]. The unsubstituted **polyRuL1** exhibits the highest efficiency (Φ_{lum}: 2.47×10^{-2})

Table 3.7 UV-vis spectral data for **polyRuL1-5** at room temperature

	LC	MLCT
	λ_{max} (nm) (ε, $M^{-1}cm^{-1}$)	λ_{max} (nm) (ε, $M^{-1}cm^{-1}$)
PolyRuL1	306 (63,700), 323 (42,200), 344 (23,700)	511 (39,600)
PolyRuL2	312 (57,200), 324 (50,700), 356 (33,000)	500 (34,900)
PolyRuL3	307 (44,300), 334 (33,500), 359 (25,600)	535 (24,200)
PolyRuL4	344 (41,900)	522 (21,700)
PolyRuL5	308 (62,300), 332 (42,400)	504 (31,300)

Solvent: 1×10^{-5} M MeOH/H_2O (v/v 4:1). The MLCT data in another paper is included in Table 3.1. λ_{max}: maximum wavelength; ε: absorption coefficient

Table 3.8 Emission data for **polyRuL1-5** at room temperature and 77 K

	λ_{max} (nm)			Φ_{lum}, × 10^{-2} (77 K)	τ (ns)	
	r.t.		77 K		r.t.	77 K
	LC band excitation	MLCT band excitation	MLCT band excitation		MLCT band excitation	MLCT band excitation
PolyRuL1	433	677	658	2.47	22.4	7900
PolyRuL2	435	661	643	1.25	12.1	5600
PolyRuL3	443	751	719	0.33	7.1	3500
PolyRuL4	445	725	704	0.21	6.6	2400
PolyRuL5	434	– [a]	650	1.09	– [a]	7400

Solvent: 5×10^{-5} M in MeOH/H_2O (v/v 4:1). [a]No detection due to weak emission. λ_{max}: maximum wavelength; Φ_{lum}: quantum yield; τ: luminescence lifetime. The Φ_{lum} is determined by comparison with the well-known laser dye, LDS750 ($\Phi_{em} = 4 \times 10^{-3}$)

among the polymers. The electron-withdrawing bromo-substituted **polyRuL5** shows at least a threefold stronger luminescence with Φ_{lum} than the electron-releasing methoxy-substituted **polyRuL3**. **PolyRuL3** shows a lower emission energy than **polyRuL1**, whereas **polyRuL5** exhibits a higher energy. The luminescence lifetimes (τ), which are measured by the time-correlated single-photon counting technique using a picosecond diode laser as the exciter, are on a nanosecond timescale at room temperature. The luminescence lifetimes at 77 K are on a microsecond timescale.

3.7 Electrochemical Switching of Emission

To electrochemically switch the emission, a solid-state device with a **polyRuL1** film is fabricated using a gel electrolyte and two transparent ITO electrodes (Fig. 3.4a) [5]. A **polyRuL1** film is prepared by spin-coating the polymer in a propanol/methanol (1:1) solution (5 g/L) on an ITO substrate. The gel electrolyte is made by mixing poly(methyl methacrylate) (PMMA), propylene carbonate, and lithium perchlorate.

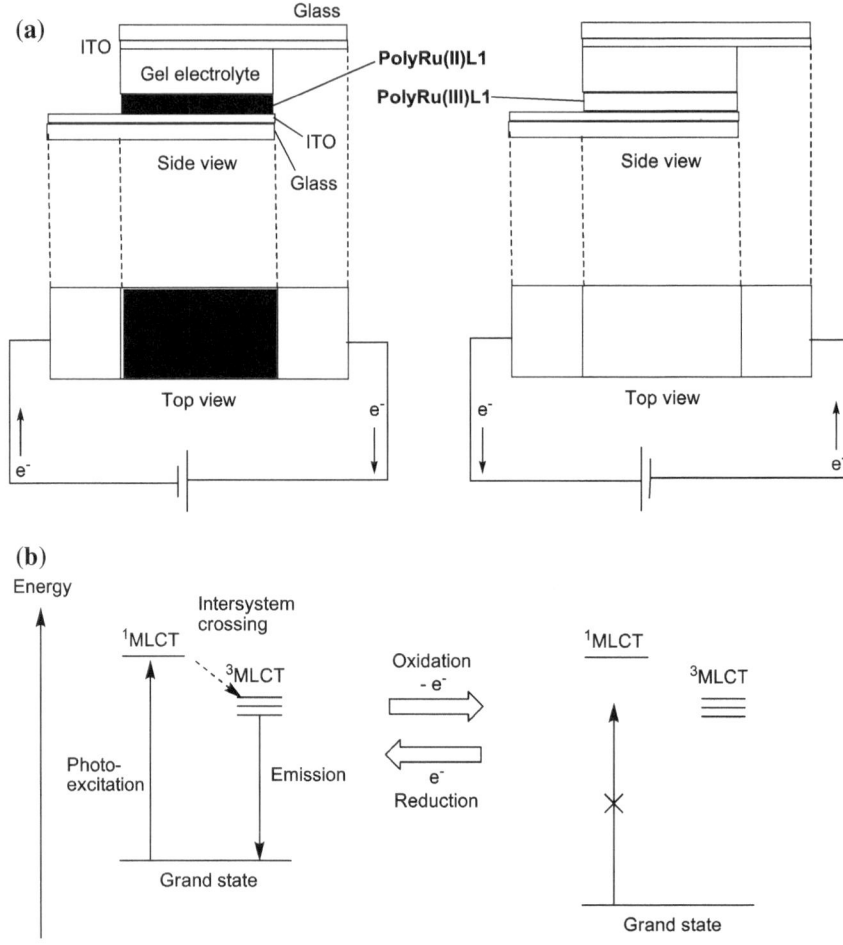

Fig. 3.4 a Colored state (left) and colorless state (right) of a solid-state device with **polyRuL1**. **b** A proposed mechanism of electrochemical switching of emissions in the device

In the device, the MLCT absorption is observed at 500 nm, and the absorption disappears by applying 2.5 V between the two ITO electrodes of the device. The colorless state is restored to the original state by applying −2.5 V. The electrochromism is caused by the electrochemical redox of Ru ions. The Ru(II) complex has a MLCT absorption, but the Ru(III) complex does not show this absorption. The original redox potential of Ru(II)/(III) in **polyRuL1** is 0.95 V versus Ag/Ag$^+$, but the application of a more oxidative potential is required to oxidize Ru(II) to Ru(III) in the device due to the internal resistance of the device. In addition, since the internal resistance is different among the devices, the voltage required to oxidize Ru(II) to Ru(III) changes for each device.

Photoluminescence in a solid-state device using **polyRuL1** is measured using microscopic spectroscopy at room temperature. When a green laser (532 nm) for the excitation of the MLCT absorption in the Ru(II) complex moieties is focused on the polymer film in the device using an objective lens, a luminescent peak at around 720 nm is observed in the photoluminescence spectrum. The emission from the polymer film is collected with an objective lens and the laser light is removed by a notch filter. The emission is detected by a Si detector with a spectrometer (all spectra are taken to account for the background and quantum efficiency of a Si detector). The luminescent peak is attributed to the emission from the ^{3}MLCT states of the Ru(II) complexes, because an electron excited by the laser light (532 nm) transits from the ^{1}MLCT band of the Ru(II) complexes to the ^{3}MLCT band via intersystem crossing and radiatively decays to the ground state (Fig. 3.4b). The intersystem crossing often occurs because of the strong spin-orbit coupling in heavy atoms such as Ru(II). In this device, the MLCT absorption disappears at 1.5 V, which means that Ru(II) ions in **polyRuL1** are totally oxidized to Ru(III) at 1.5 V. The emission also disappears at this voltage and can be almost restored to its original intensity by applying the reverse voltage (-1.5 V). The quenched and emitted states are reversibly switched by applying the two voltages alternately. The quenching is likely caused by the disappearance of the MLCT absorption based on the oxidation of Ru(II) to Ru(III).

References

1. Han FS, Higuchi M, Kurth DG (2008) Metallo-supramolecular polyelectrolytes self-assembled from various pyridine ring substituted bis-terpyridines and metal ions: photophysical, electrochemical and electrochromic properties. J Am Chem Soc 130:2073–2081. https://doi.org/10.1021/ja710380a
2. Hu CW, Sato T, Zhang J, Moriyama T, Higuchi M (2013) Multi-colour electrochromic properties of Fe/Ru-based bimetallo-supramolecular polymers. J Mater Chem C 1:3408–3413. https://doi.org/10.1039/c3tc30440j
3. Chen BH, Kao SY, Hu CW, Higuchi M, Ho KC, Liao YC (2015) Printed multicolor high-contrast electrochromic devices. ACS Appl Mater Interfaces 7:25069–25076. https://doi.org/10.1021/acsami.5b08061
4. Han FS, Higuchi M, Ikeda T, Negishi Y, Tsukuda T, Kurth DG (2008) Luminescence properties of metallo-supramolecular coordination polymers assembled from pyridine ring functionalized ditopic bis-terpyridines and Ru(II) ion. J Mater Chem 18:4555–4560. https://doi.org/10.1039/b806930a
5. Muronoi Y, Zhang J, Higuchi M, Maki H (2013) Electrochemical switching of luminescence in Ru(II)-based metallosupramolecular polymer device. Chem Lett 18:4555–4560. https://doi.org/10.1246/cl.130158

Chapter 4
Co(II)-Based Metallo-Supramolecular Polymers

4.1 Synthesis

A Co(II) ion has six coordination sites with an octahedral structure, like Fe(II) and Ru(II) ions. Therefore, it forms a stable 1:2 complex with a terpyridine moiety. Co(II)-based metallo-supramolecular polymers (**polyCoL1-5**) are synthesized by the 1:1 complexation of Co(OAc)$_2$ and bis(terpyridine)s (**L1-5**) (Fig. 4.1) [1]. The color of the polymers is orange or yellow. The typical synthetic procedure is as follows. Equimolar amounts of **L1** and Co(OAc)$_2$ are refluxed in methanol under argon atmosphere for 24 h. The resulting solution is cooled to room temperature and filtered to remove a small amount of insoluble residues. The filtrate is slowly evaporated to remove the solvent, and the product is collected and dried further in vacuum overnight to give **polyCoL1** (>90%). In gel permeation chromatography (GPC), it is generally believed that there is no chemical interaction between an injected sample and the column filler. Therefore, molecular weight of the sample can be estimated exactly from only the elution time. However, metallo-supramolecular polymers such as **polyCoL1** are dissociated inside the column and it is impossible to determine the molecular weight. This is probably because very weak interaction actually existing between the polymer and the column filler breaks the coordination bonds of polymer.

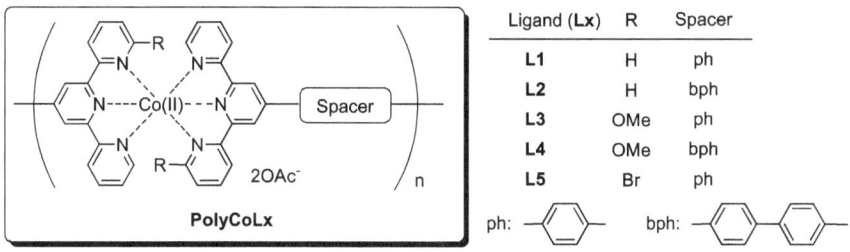

Ligand (**Lx**)	R	Spacer
L1	H	ph
L2	H	bph
L3	OMe	ph
L4	OMe	bph
L5	Br	ph

Fig. 4.1 Co(II)-based metallo-supramolecular polymers (**polyCoL1-5**)

M. Higuchi, *Metallo-Supramolecular Polymers*, NIMS Monographs,
https://doi.org/10.1007/978-4-431-56891-9_4

Recently, it became obvious that addition of a salt such as LiBr in the eluent is effective to weaken the interaction between the polymer and the column filler [2]. The molecular weight (M_w) of **polyCoL1** is determined to be 4.31×10^3 using GPC by the addition of a salt to the eluent to prevent decomplexation (Instrument: LC-9110 NEXT, JAI; column: JAIGEL-3HAF+4HAF; eluent: 100 mM LiBr in DMF; flow rate: 1.0 mL/min; detector: UV-370 NEXT@268 nm; injection volume: 0.1 mL; standard: polystyrene). The polymer is dissolved in methanol (4 mg/mL). In solutions of high concentration, the polymers aggregate, probably due to strong ionic interactions among cationic metallo-polymer chains through the counter anions. Once the **polyCoL1** film has formed, it becomes insoluble in water, and strong treatment such as ultra-sonification is necessary to dissolve the polymer film.

4.2 Yellow Electrochromism

In the UV-vis spectra of **polyCoL1-5**, a weak absorption based on the d-d* transition of Co(II) appears around 520 nm (Table 4.1). MLCT absorptions are not observed in these polymers. The absorption coefficient (ε_{max}) of the absorption based on the d-d* transition is markedly reduced by the introduction of methoxy and bromo groups onto the ligand. The redox potential of Co(II)/(III) in the polymers was measured by cyclic voltammetry, but the redox wave was very small. The potential was successfully determined by differential pulse voltammetry (DPV) for **polyCoL2**. Reliable data was not obtained for **polyCoL3-5**. Regarding the electrochromic properties, the disappearance/reappearance of the absorption due to the d-d* transition was observed in **polyCoL1-2** by applying 0.5 or −0.5 V versus Ag/Ag+, respectively. This color change from yellow to colorless was observed. **PolyCoL3-5** do not show electrochromism due to the very small absorption of the d-d* transition.

Table 4.1 Optical and electrochemical properties of Co(II)-based metallo-supramolecular polymers (**polyCoL1-5**)

	Maximum wavelength (λ_{max})[a] (nm)	Absorption coefficient (ε)[a]	Redox potential ($E_{1/2}$)[b], V versus Ag/Ag+
PolyCoL1	522	2514	0.02
PolyCoL2	521	2300	0.04[c]
PolyCoL3	521	214	−[d]
PolyCoL4	519	457	−
PolyCoL5	526	429	−

[a]Solvent: methanol. [b]Working electrode: glassy carbon; counter electrode: Pt wire; reference electrode: Ag/Ag+; electrolyte: 0.10 M n-Bu$_4$NClO$_4$/acetonitrile; scan rate: 100 mV/s. [c]Measured by differential pulse voltammetry (DPV). [d]Cannot be determined

4.3 Black Electrochromism

Blue or red electrochromism is achieved by the electrochemical redox between Fe(II)/(III) or Ru(II)/(III) ions in Fe(II)- or Ru(II)-based metallo-supramolecular polymer films, respectively. Yellow electrochromism is also realized using the redox between Co(II)/(III) ions in the polymer film in an organic solvent including an electrolyte. However, in general, black electrochromism is difficult to achieve because a broad absorbance covering the visible light region is required. We accidentally observed black electrochromism in **polyCoL1** in an aqueous solution including an electrolyte [2]. The redox between Co(I)/(II) is used in this black electrochromism, and the Co(I) complex state is stabilized by water molecules.

A **polyCoL1** film is prepared by spray-coating the methanol solution onto an ITO glass with a mask to confine the active area to 1×3 cm^2. The film thickness is measured as approximately 800 nm by scanning electron microscopy (SEM). The polymer film is stable in water: no polymer dissolution is observed even after continuous immersion for more than 30 days. In the cyclic voltammogram (CV) of the polymer film in 0.1 M aqueous KCl solution, a large reversible redox wave of Co(I)/(II) is observed at -0.86 V versus Ag/Ag$^+$ ($E_{1/2}$). The redox of Co(II)/(III) is also observed at 0.16 V versus Ag/Ag$^+$ ($E_{1/2}$) in the CV, but the current is much smaller than that of Co(I)/(II), which is attributed to different self-electron-exchange abilities. The self-exchange electron transfer rate constant of Co(II)/(III) (2 M^{-1} S^{-1}) has been reported to be much smaller than that of Co(I)/(II) (10^8 M^{-1} S^{-1}). In situ UV–vis–NIR spectra of the polymer film reveal the appearance of a very broad absorption between 500 and 1700 nm when a more negative potential than -0.7 V versus Ag/Ag$^+$ is applied. This black electrochromism is caused by the electrochemical reduction from Co(II) to Co(I). Since Co(I) is a richer electron donor than Co(II), the increasing extent of splitting of the d orbitals in the metal may result in stronger d–d* transitions that also experience a redshift from 520 to 560 nm. In addition, a highly intense absorption with a maximum wavelength around 1500 nm is observed in the near-infrared (NIR) region. This absorption is attributed to the MLCT between the stronger electron donor ability of Co(I) ions and the good electron acceptor ability of the terpyridine moieties. The **polyCoL1** film in its reduced state exhibits a broad absorption covering the UV–vis–NIR regions, which, to the best of our knowledge, is rare in electrochromic materials.

The orange color of the polymer dissolved in water becomes lighter by the addition of a base. When NaOH is added to an aqueous solution of **polyCoL1**, a significant decrease in the intensity of the d–d* absorption around 520 nm is observed in the UV–vis spectra. In fact, the partially filled d subshell of the Co ion leads to its high activity in ligand complexes or catalysis, and thus it is sensitive to the presence of the hydroxide ions. On the basis of these findings, the electrochromic properties of a **polyCoL1** film were investigated in solutions at different pHs (Table 4.2). Higher transparency in the bleached state of the polymer film during electrochromic change was achieved in an alkaline solution ($T_{bleached}$ at 550 nm: 20.6% in pH 7 and 81.9% in pH 13). ΔT reaches 74.3% at pH 13. A switching test between 0.0 and -1.3 V

Table 4.2 Electrochromic properties of a **polyCoL1** film in solutions of different pHs

pH	$T_{bleached}$ (%)	$T_{colored}$ (%)	ΔT (%)	$T_{bleaching}$ (s)	$T_{coloring}$ (s)
7	20.6	2.5	18.1	3.3	6.0
9	23.4	3.6	19.8	4.6	21.9
11	54.3	5.0	49.3	11.1	28.8
13	81.9	7.6	74.3	23.6	31.5

The transmittances of the MLCT absorption (λ_{max}: 550 nm) in the bleached ($T_{bleached}$) and colored states ($T_{colored}$) of polymer films coated on ITO glass are measured by in situ UV–vis spectroscopy at 0 or -1.3 V versus Ag/AgCl/Sat'd KCl. The transmittance difference (ΔT) is calculated from $T_{bleached}$ and $T_{colored}$. The times for coloring and bleaching ($t_{coloring}$ and $t_{bleaching}$) are defined as the time taken for 95% of ΔT to change

in 100 steps demonstrated that the polymer film had good electrochromic durability, and the transmittance change remained within 94.5% of its initial value. However, the response times became longer as the basicity of the solution was increased (t_b: 23. 6 s; t_d: 31.5 s in pH 13) in comparison with those in a solution at pH 7 (t_b: 3.3 s; t_d: 6.0 s), probably because low electron transfer ability is associated with increasing concentrations of hydroxide ions.

4.4 Polymer Formation via Polycondensation Between the Ligand Moieties

Since metallo-supramolecular polymer chains are composed of coordinate covalent bonds, the simplest way to synthesize the polymers is to use complexation reactions of metal ions and ditopic ligands, as described in the previous sections, such as the complexation of cobalt ions with bis(terpyridine)s. However, the other synthetic methods have been considered to obtain metallo-supramolecular polymers. For example, metallo-supramolecular polymers can be prepared by A_2/B_2 polycondensation between the ligand moieties of the 1:2 complexes bearing two reaction sites (A_2 monomer) with another B_2 monomer (Fig. 4.2a) [3]. One of the requirements for this type of polymerization is that no decomposition of the 1:2 complex occurs during the polycondensation between the ligand moieties.

A 1:2 complex of Co(III) ion with a tridentate azo ligand bearing an amino group as the reactive group (Fig. 4.2b) was synthesized according to the literature. The oxidation state of Co is 3+, the azo ligand is mono-anionic, and one perchlorate anion exists in the complex. The metallo-supramolecular polymer with a π-conjugated phenyl group as the spacer, **polyCoL7**, was prepared by the polycondensation of the 1:2 complex with terephthalaldehyde in dehydrated ethanol in the presence of Si(OEt)$_4$ as a dehydrating agent. The polymer was obtained as a bright-green solid in 77% yield and was soluble in common organic solvents except for methanol. The ^{1}H NMR spectrum of **polyCoL7** shows peaks at $\delta 10.003$ and $\delta 5.73$ corresponding

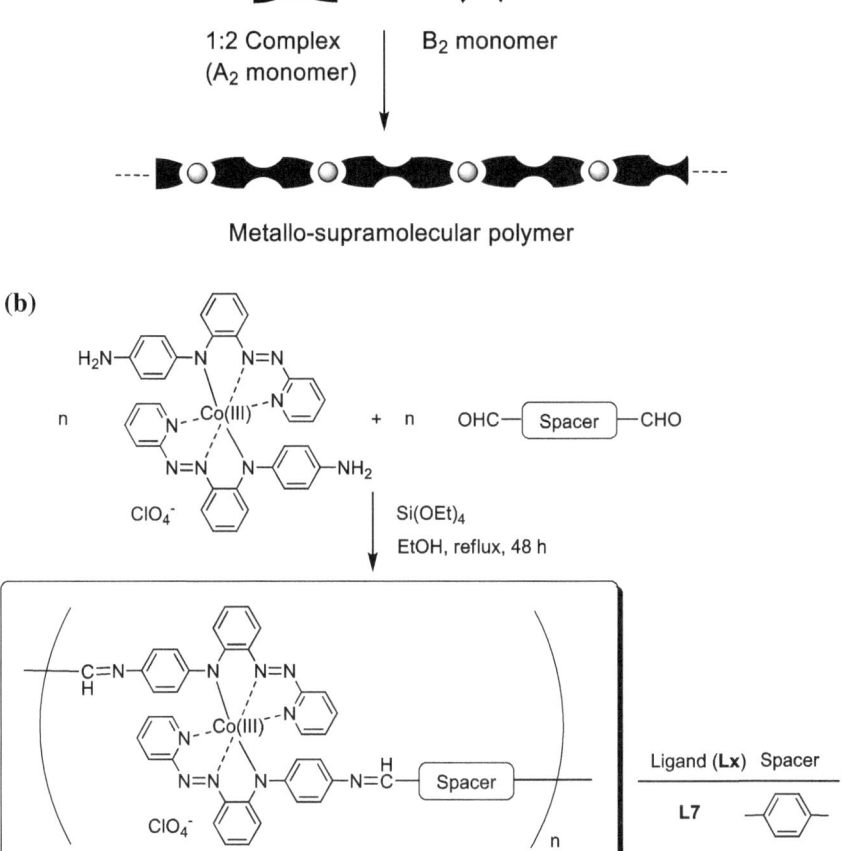

Fig. 4.2 a Metallo-Supramolecular polymer formation by polycondensation of the 1:2 complex bearing two reaction sites (A₂ monomer) and B₂ monomer. **b** Synthesis of Co(II)-based metallo-supramolecular polymers (**polyCoL7-8**)

to the terminal aldehyde proton and the terminal amine proton, respectively. The integration of the two terminal peaks is almost 1:1, which supports the fact that the Co monomer and terephthalaldehyde were added at a 1:1 ratio in the polymerization. The degree of polymerization (DP) of **polyCoL7** is estimated to be 18 from the ratio of the integration of the terminal proton peak to the internal proton peak. **PolyCoL8** with a nonconjugated spacer is also synthesized by the same procedure using glutaraldehyde.

4.5 Nonvolatile Memory

To the three conventional electronic devices (resistor, capacitor, and inductor), Leon O. Chua proposed a new fourth circuit element, a memristor, that could process information similar to biological systems. Recently, this conceptual device has been explored in a doped titanium-dioxide matrix. It has been demonstrated that hysteresis in the current–voltage (I–V) characteristic has the potential to emulate a neural network. Exploring an intriguing avenue of atomic engineering is required to further advance the memristor architecture.

Memristor devices made with **polyCoL7-8** are fabricated as follows. Acetonitrile and DMSO solutions of **polyCoL7** and **polyCoL8**, respectively, are dropped onto a Si/SiO_2 substrate and dried. The substrates had interdigitated Au electrodes (lateral gap ~200 nm) fabricated by e-beam lithography. A current lateral structure is chosen for the device (Fig. 4.3). To prevent an artifact induced by the pinning effect during comparison, a sandwich structure is avoided, even though it would certainly provide superior performance to the lateral structure. The measurements are performed under ambient atmospheric conditions using an Agilent 4155c semiconductor parameter analyzer attached to a Vector probe station. The I–V characteristics of the ligand alone do not show any switching up to ±16 V. Therefore, the switching phenomenon in the polymer devices is based on the reduction of Co(III) to Co(II).

Molecular analogs of the memristive matrices used here are electrochemically active conjugated Co(III) polymer (**polyCoL7**) and a nonconjugated Co(III) polymer (**polyCoL8**). The redox switching in the metallo-supramolecular polymers generates bistable states with a large ON/OFF ratio that supports random flip-flops for several hours. Thus, the results provide a synthetic solution to leakage current restriction, one of the fundamental problems faced when fabricating state-of-the-art electronic devices.

Fig. 4.3 A memristor device structure with **polyCoL7**

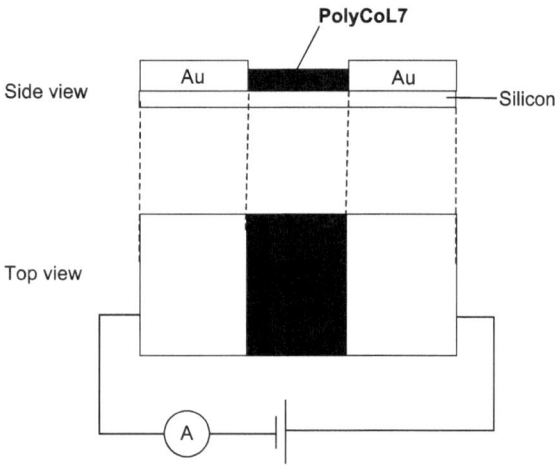

The RAM and ROM properties of a **polyCoL7** device enable reversible information processing and permanent memory storage capacities. When a write $(-5\,\mathrm{V})$-read $(-1\,\mathrm{V})$-erase $(+5\,\mathrm{V})$-read $(-1\,\mathrm{V})$ cycle is sent through the device to test the RAM, it is possible to carry out RAM/ROM operations at ± 3, ± 4, ± 5, ± 6, ± 7, ± 8, ± 9, ± 10, ± 11, ± 12, ± 13, ± 14, and ± 15 V. Therefore, the **polyCoL7** device shows remarkable multilevel switching that can potentially replace the existing binary computing devices with a higher-logic system. The ON/OFF ratio for the **polyCoL7** device is on the order of 10^2, whereas that for **polyCoL8** is <10. The ROM feature offered several read currents (70 s) after writing a state once (solid circle), and then the same process was repeated after erasing once (hollow circle). The **polyCoL7** device was probed for more than 3 h, and it showed little degradation in its high (ON) and low (OFF) conducting states. The **polyCoL7** device showed unusually high robustness in air, and its retention time was considerably longer than that of the **polyCoL8** device. Since no external signal was applied to the device when the ROM effect was being tested, the rapid weakening of the ROM effect in the **polyCoL8** matrix clearly indicated that the spontaneous destruction of memory is due to the large leakage current generated by phonons and the subsequent isolation of clusters. Generally, it is considered that an increase in the matrix conductivity would lead to an increase in the OFF (leakage) and ON currents. The result is just opposite in molecular engineering. In other words, molecular engineering helps to block the leakage current and channel it to stabilize the ON and OFF states. Therefore, instead of metal ion doping in conducting polymers or inorganic semiconductors, if the metal ion is bonded directly to the conjugated organic moiety, then leakage current tuning would enhance memristive features significantly.

References

1. Han FS, Higuchi M, Kurth DG (2008) Metallo-supramolecular polyelectrolytes self-assembled from various pyridine ring substituted bis-terpyridines and metal ions: photophysical, electrochemical and electrochromic properties. J Am Chem Soc 130:2073–2081. https://doi.org/10.1021/ja710380a
2. Hsu CY, Zhang J, Sato T, Moriyama S, Higuchi M (2015) Black-to-transmissive electrochromism with visible-to-near-infrared switching of a Co(II)-based metallo-supramolecular polymer for smart window and digital signage applications. ACS Appl Mater Interfaces 7:18266–18272. https://doi.org/10.1021/acsami.5b02990
3. Bandyopadhyay A, Sahu S, Higuchi M (2011) Tuning of nonvolatile bipolar memristive switching in Co(III) polymer with an extended azo aromatic ligand. J Am Chem Soc 133:1168–1171. https://doi.org/10.1021/ja106945v

Chapter 5
Zn(II)-Based Metallo-Supramolecular Polymers

5.1 Synthesis

Zn(II) ions can be introduced to a metallo-supramolecular polymer backbone using bis(terpyridine)s because of the octahedral geometry of the six-coordinate Zn(II) ion (Fig. 5.1) [1]. The Zn(II)-based metallo-supramolecular polymers (**polyZnL2,4,9**) are synthesized as follows. Equimolar amounts of bis(terpyridine) (**L2, L4**, or **L9**) and Zn(ClO$_4$)$_2$ hexahydrate are stirred at 80 °C in argon-saturated N-methylpyrroridone (NMP) (ca. 1 mL of solvent per mg of bis(terpyridine)) for 24 h. After the solution is cooled to room temperature, diethyl ether is added slowly until the solution is colorless. The precipitated polymers are collected by filtration and washed with diethyl ether three times, then dried *in vacuo* overnight to give polymers with a 99% (**polyZnL2**), 75% (**polyZnL4**), or 87% (**polyZnL9**) yield. Molecular weight of the polymers couldn't be determined by GPC measurement because of the decomplexation in the GPC column.

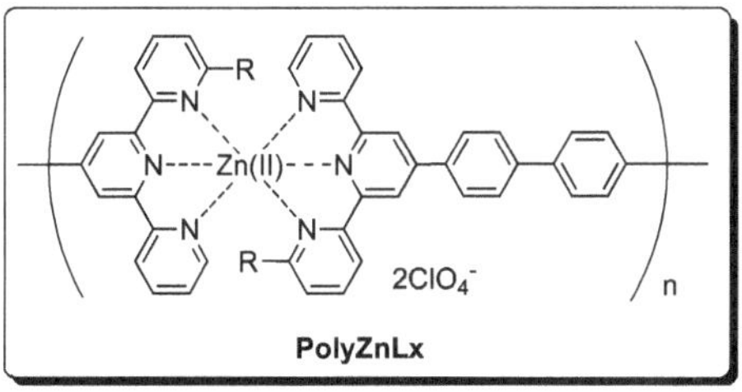

Ligand (Lx)	R
L2	H
L4	OMe
L9	CN

PolyZnLx

Fig. 5.1 Zn(II)-based metallo-supramolecular polymers (**polyZnL2,4,9**)

5.2   Fluorescent Color Modulation

Zn(II)-based metallo-supramolecular polymers have been extensively studied with regard to their prominent photoluminescent (PL) and electroluminescent (EL) properties. Although metal-to-ligand charge transfer (MLCT) does not occur due to the filled electron d^{10} shell in Zn(II) ions, the Zn(II)-based polymers show intraligand charge transfer (ILCT) between the coordination site and the chromophore of the ligand. Therefore, the emissive nature of the polymers greatly depends on the ligand structure. Zn(II)-based polymers with various bis(terpyridine) ligands bearing a different chromophore as the spacer unit have been reported so far. Würthner et al. synthesized the Zn(II)-based polymers with bis(terpyridine) bearing a perylene bis(imide) spacer. Che et al. reported that Zn(II)-based polymers with bis(terpyridine)s bearing π-conjugated spacers exhibited high PL quantum yields and a wide range of emission colors from violet to yellow [2]. Lin et al. also reported emission properties of Zn(II)-based polymers with fluorene derivatives as the spacer [3, 4]. Furthermore, Schubert et al. reported the synthesis of Zn(II)-based polymers with π-conjugated spacers bearing electron-donating or electron-accepting groups, their photophysical properties, and their utility in sensing cyanide and phosphates [5]. The modification of the ligands was mainly done at the 4'-position of the central pyridine of the terpyridine moiety. Studies on functionalization at the 6- and/or 6''-positions of the peripheral pyridines are few to the best of our knowledge. Herein, synthesis of Zn(II)-based metallo-supramolecular polymers with bis(terpyridine) substituted at the 6-/6''-positions of the peripheral pyridine and their fluorescent properties are described. The amorphous metallo-supramolecular polymers with bis(terpyridine)s as the ligand have made homogeneous film formation possible. UV-vis spectra of **polyZnL2,4,9** resemble those of the corresponding ligands (**L2**, **L4**, and **L9**) (Table 5.1), because Zn(II) ions with a filled d-shell do not participate in the electron transitions. The polymers show strong ligand-centered (LC) π-π* or n-π* transitions

Table 5.1   UV–vis spectral data of **polyZnL2,4,9** and the ligands

	In solution		Film
	λ_{abs} (nm)	$\varepsilon \times 10^4$ (M^{-1} cm^{-1})	λ_{abs} (nm)
PolyZnL2	290, 315	4.40	290, 353
PolyZnL4	315	7.28	355
PolyZnL9	288, 315	4.48	290, 337
L2	255, 301, 313	5.82	–
L4	254, 313	7.56	–
L9	287, 309	4.22	–

PolyZnL2,4,9: 5×10^{-5} M in DMF. **L2,4,9**: 5×10^{-5} M in CHCl$_3$. The absorption coefficients (ε) are those of the lowest-energy absorption band. Thin films of **polyZnL2,4,9** are prepared on a quartz glass substrate by spin-coating from DMF solutions (1 mg/mL, 300 rpm for 500 s, 1000 rpm for 500 s). The thicknesses of the **polyZnL2,4,9** films were measured using a micrometer to be about 30, 10, and 30 nm, respectively

around 315 nm, which is red-shifted about 2–6 nm compared with the corresponding ligand, probably due to the inductive effect of the metal ions. **PolyZnL4** with electron-donating groups in the ligand displays a significantly increased intensity of absorption around 315 nm, compared with the unsubstituted **polyZnL2**, owing to the red shift of the peak around 280 nm because of the electron-donating effect. **PolyZnL9** with electron-withdrawing groups showed a blue shift of the peak around 280 nm in comparison with **polyZnL2**. In the UV-vis spectra of the **polyZnL2,4,9** films, absorption peaks are red-shifted about 20–40 nm compared with those of **polyZnL2,4,9** in solution. Such a red shift is commonly observed in π-conjugated polymers, because the intermolecular π-π stacking between the polymer chains is strong in the solid state. On the other hand, the powder XRD spectrum of **polyZnL2** showed only broad peaks based on the amorphous polymer structure. This result indicates that various π-π stacking structures exist among the polymer chains, probably because the bulky Zn(II) complex moieties with the octahedral geometry prevent the formation of a specific π-π stacking structure between the polymer chains.

The photoluminescence properties of **polyZnL2,4,9** and the ligands in solution were investigated by exciting the absorption maxima of the lowest-energy band (Table 5.2). The maximum wavelength (λ_{PL}) of the emission peak in dilute DMF solutions of the polymers were in the range of 375–383 nm. These bonds are assigned to the emission from the ligand-centered (LC) π-π* and n-π* transitions. The λ_{PL}s of the polymers are similar to those of the corresponding ligands, but the peaks in the polymers are broader than those in the ligands due to complexation. When the absolute PL quantum yield (Φ_{PL}) in solution is compared between **polyZnL2,4,9** and **L2,4,9**, the Φ_{PL}s of the polymers ($\Phi_{PL} = 0.69–0.76$) are lower than those of the ligands ($\Phi_{PL} = 0.80–0.93$). A similar decrease in Φ_{PL} in the polymers has often been reported for other Zn(II)-based metallo-supramolecular polymers. The λ_{abs}s in **polyZnL2,4,9** films are red-shifted about 20–40 nm compared with those in solution. Moreover, the shift of λ_{PL} in the polymer films is very large (63, 93, and 144 nm, respectively). In particular, the photoluminescence of the **polyZnL9** film shows the largest Stokes shift. It is proposed that the dipole moment change in the ligand caused

Table 5.2 Emission data for **polyZnL2,4,9** and the ligands

	In solution			Film		
	λ_{PL} (nm)	Stokes shift (cm^{-1})	Φ_{PL}	λ_{PL} (nm)	Stokes shift (cm^{-1})	Φ_{PL}
PolyZnL2	379	5360	0.69	442	5700	0.24
PolyZnL4	375	5080	0.73	468	6800	0.25
PolyZnL9	383	5640	0.68	527	10,700	0.19
L2	369, 380	5630	0.93	–	–	–
L4	368, 375	5280	0.87	–	–	–
L9	373, 382	6180	0.91	–	–	–

PolyZnL2,4,9: 2.5×10^{-5} M in DMF. **L2,4,9**: 2.5×10^{-5} M in CHCl₃. Absolute quantum yields, uncorrected with respect to reabsorption

by the electron-withdrawing cyano groups promoted interchain π-π staking in the polymer film, which resulted in a large extension of π-conjugation. The lower Φ_{PL} of the **polyZnL9** film ($\Phi_{PL} = 0.19$) compared with that of **polyZnL2** ($\Phi_{PL} = 0.24$) also indicates an increase in the intermolecular π-π stacking between the polymer chains. A large Stokes shift is observed in the **polyZnL4** film as well. It is considered that the electron-donating effect of the methoxy groups also enhanced the intermolecular π-π stacking. The Zn(II)-based polymer films showed vastly different luminescence colors, ranging from blue to cyan to green, caused by the introduction of electron-donating/withdrawing groups to the 6-positions of the bis(terpyridine) ligands. In the CIE system, the luminescent colors of the **polyZnL2,4,9** films are blue (x: 0.18, y: 0.19), cyan (x: 0.24, y: 0.30), and green (x: 0.31, y: 0.45), respectively. A Stokes shift between the film and the solution states has been observed in the other Zn(II)-based metallo-supramolecular polymers, but such a Stokes shift (144 nm) as large as that observed in **polyZnL9** has not been reported so far.

References

1. Sato T, Pandey RK, Higuchi M (2013) Fluorescent colour modulation in Zn(II)-based metallosupramolecular polymer films by electronic-state control of the ligand. Dalton Trans 42:16036–16042. https://doi.org/10.1039/c3dt51354h
2. Yu SC, Kwok CC, Chan WK, Che CM (2003) Self-assembled electroluminescent polymers derived from terpyridine-based moieties. Adv Mater 15:1643–1647. https://doi.org/10.1002/adma.2003050002
3. Chen YY, Lin HC (2007) Synthesis and characterization of light-emitting main-chain metallopolymers containing bis-terpyridyl ligands with various lateral substituents. J Polym Sci, Part A: Polym Chem 45:3243–3255. https://doi.org/10.1002/pola.22073
4. Chen YY, Tao YT, Lin HC (2006) Novel self-assembled metallo-homopolymers and metallo-alt-copolymer containing terpyridyl zinc(II) moieties. Macromolecules 39:8559–8566. https://doi.org/10.1021/ma0618629
5. Wild A, Teichler A, Ho CL, Wang XZ, Zhan H, Schlutter F, Winter A, Hager MD, Wong WY, Schubert US (2013) Formation of dynamic metallo-copolymers by inkjet printing: towards white-emitting materials. J Mater Chem C 1:1812–1822. https://doi.org/10.1039/c2tc00552b

Chapter 6
Cu(II)-Based Metallo-Supramolecular Polymers

6.1 Synthesis

To introduce Cu(II) ions with a tetrahedral structure into metallo-supramolecular polymers, 5,5′-linked bis(1,10-phenanthroline)s (**L10-15**) were prepared [1, 2]. A Cu(II)-based metallo-supramolecular polymer, **polyCuL13**, is prepared by the 1:1 complexation of **L13** and $Cu(ClO_4)_2 \cdot 6H_2O$ (Fig. 6.1).

L11 was synthesized from 5-bromo-1,10-phenanthroline and 1,4-benzenediboronic acid bis(pinacol) ester by a Suzuki-type cross-coupling reaction. Purification of the reaction mixture by alumina column chromatography and preparative HPLC affords the product **L11**. The yield of ligand **L11** highly depends on the synthetic conditions (Table 6.1). $Pd(PPh_3)_4$ is also useful as a catalyst, but $PdCl_2(PPh_3)_2$ is easier to handle and more stable against oxygen than $Pd(PPh_3)_4$ (entries 1, 2). When KOAc is used instead of K_2CO_3, the yield of **L11** decreases owing to the lower basicity (entry 3). DMSO is the best solvent among dioxane, THF, and DMSO because of the extremely poor solubility of **L11** in dioxane and THF (entries 4, 5). Although the amount of catalyst is increased from 5 to 10 mol%, the yield does not improve and actually decreases owing to the formation of byproducts (entry 6). The reaction run on a tenfold larger scale yielded almost the same amount of product (entry 7). The other ligands (**L10, 12-15**) were prepared under similar conditions to those in entry 1.

The detailed synthetic procedure for **PolyCuL13** is as follows. Under a nitrogen atmosphere, **L13** (19.6 mg, 0.026 mmol) is dissolved in acetonitrile (5 mL). $Cu(ClO_4)_2 \cdot 6H_2O$ (9.7 mg, 0.026 mmol) dissolved in acetonitrile (5 mL) is added dropwise over 30 min to the ligand solution. The mixture is stirred for 1 h at room temperature. The solvent is slowly evaporated from the green solution by passing a stream of nitrogen over the reaction mixture. The green precipitate is rinsed thoroughly with diethyl ether (20 mL), filtered, and dried in vacuo at room temperature. Yield: 26.4 mg, 90%. Molecular weight of the polymer couldn't be determined by GPC measurement because of the decomplexation in the GPC column.

© National Institute for Materials Science, Japan 2019
M. Higuchi, *Metallo-Supramolecular Polymers*, NIMS Monographs,
https://doi.org/10.1007/978-4-431-56891-9_6

(a)

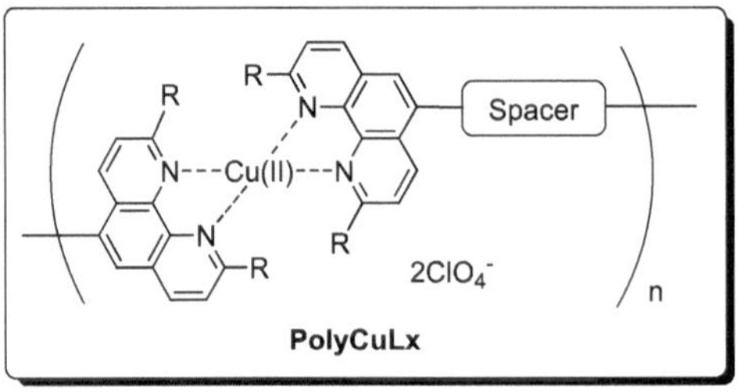

Ligand (**Lx**)	R	Spacer
L10	Me	ph
L11	H	ph
L12	Me	fl
L13	H	fl
L14	Me	none
L15	H	none

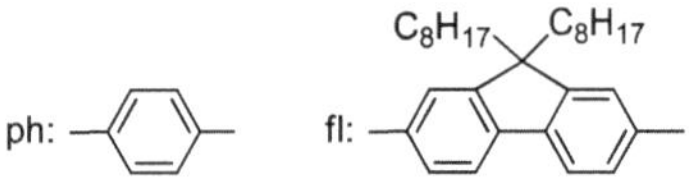

(b)

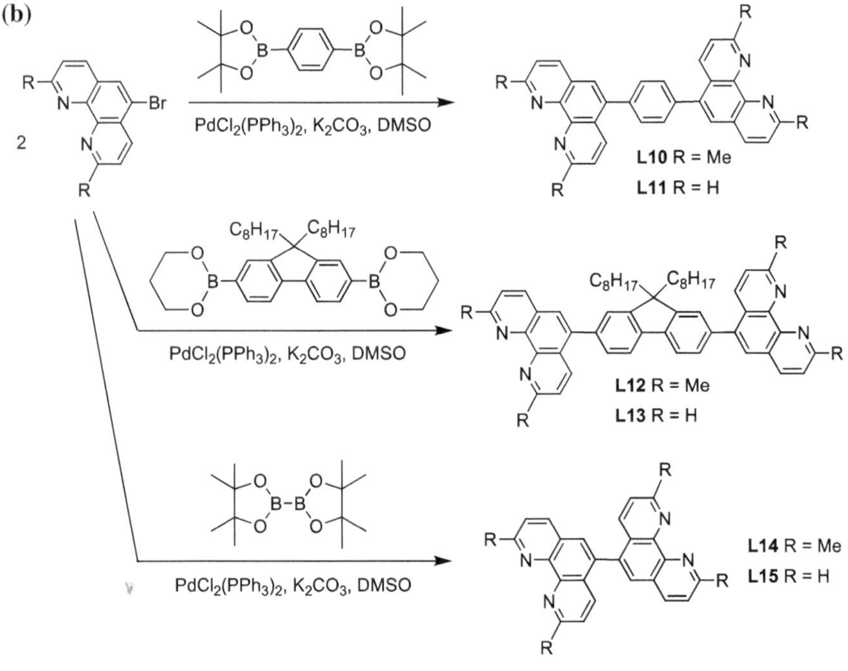

Fig. 6.1 **a** Cu(II)-based metallo-supramolecular polymers (**polyCuL10-15**) and **b** the synthesis of **L10-15**

The polymer structure of **polyCuL13** was observed by atomic force microscopy (AFM) under ambient conditions. The polymer (3 μM in acetonitrile) spin-coated on a freshly cleaved highly oriented pyrolytic graphite (HOPG) surface shows filamentous structures longer than 15 μm. From the comparison of the chain height in the AFM image with the calculated width of the single polymer chain, the long filamentous structures are considered to be a mixture of single polymer chains and bundles of several chains.

Table 6.1 Synthesis of **L11** by a Suzuki coupling reaction of 5-bromo-1,10-phenanthroline with 1,4-benzenedibromic acid bis(pinacol) ether under different conditions

Entry	Catalyst	Base	Solvent	Temp. (°C)	Yield (%)
1.	$PdCl_2(PPh_3)_2$	K_2CO_3	DMSO	100	75
2.	$Pd(PPh_3)_2$	K_2CO_3	DMSO	100	65
3.	$PdCl_2(PPh_3)_2$	KOAc	DMSO	100	40
4.	$PdCl_2(PPh_3)_2$	K_2CO_3	1,4-dioxane	100	10
5.	$PdCl_2(PPh_3)_2$	K_2CO_3	THF	65	30
6[a].	$PdCl_2(PPh_3)_2$	K_2CO_3	DMSO	100	70
7[b].	$PdCl_2(PPh_3)_2$	K_2CO_3	DMSO	100	72

Conditions for entries 1–5: 5-bromophenanthroline (0.2 g, 0.8 mmol), 1,4-benzenedibromic acid bis(pinacol) ether (0.4 mmol), catalyst (5 mol%), base (2.4 mmol), solvent (30 mL), 24 h under nitrogen atmosphere. [a]Catalyst (10 mol%). [b]solvent (250 mL), 30 h

6.2 Green Electrochromism

The complexation of **L13** with Cu(II) salt was confirmed through UV–vis spectrophotometric titration. The ligand-centered absorption (λ_{max}: 324 nm) was observed in the spectrum of an acetonitrile solution of **L13** at room temperature. When $Cu(ClO_4)_2 \cdot 6H_2O$ was added to the ligand solution, a red shift of the absorption to 336 nm was observed. At the same time, a shift of the isosbestic points was also observed during the titration, indicating that the complexation proceeds in a stepwise fashion. An isosbestic point appears when a compound is quantitatively transformed to another one during a titration. Therefore, a shift of the isosbestic points means that two kinds of complexation occur stepwise. From the result that the shift proceeds when 0.5 equiv. of Cu(II) ion is added to **L13**, the 2:1 complex forms first during the titration, and then the polymerization takes place by the complexation of the 2:1 complex with the additional Cu(II) ions (Fig. 6.2). This result indicates that the complexation constant for the first complexation (K_1) is higher than that for the second (K_2), probably because the electron-donating ability of the phenanthroline moieties in **L13** decreases when the 2:1 complex forms with the cationic Cu(II) ion.

In general, Cu(II) ions form a tetrahedral structure with bidentate ligands such as 1,10-phenanthroline or bipyridine ligands, and the Cu(II) complexes are known to have a green color. The synthesized polymer, **polyCuL13**, also is green. The reflection peak in **polyCuL13** appears at 524 nm, which is in the green range (480–560 nm).

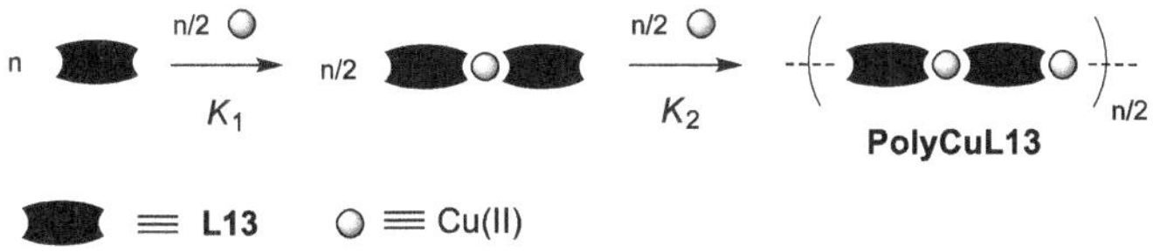

Fig. 6.2 a Stepwise complexation behavior of **L13** with Cu(II) ions during the titration

The green color is also confirmed in the CIE 1931 chromaticity diagram ($x = 0.236$, $y = 0.675$). A redox wave of Cu(II)/(I) is observed around -0.8 V versus Ag/Ag$^+$ in the cyclic voltammogram of a **polyCuL13** film (working electrode: glassy carbon; counter electrode: Pt wire; electrolyte: 0.1 M TBAP/acetonitrile; scan rate: 100 mV/s). However, the polymer film is gradually dissolved by the electrolyte solution during the measurement. Therefore, the electrochromic properties are evaluated in solution. When -1.2 V versus Ag/Ag$^+$ is applied to the solution, the green color disappears. The colorless state changes to green again by applying 1.2 V versus Ag/Ag$^+$. The green color is based on the MLCT absorption between Cu(II) and the phenanthroline moiety in **L13**. Therefore, the disappearance of the MLCT absorption is caused by the electrochemical reduction of Cu(II) to Cu(I).

6.3   Cu(I)/Fe(II)-Based Heterometallo-Supramolecular Polymer

A Cu(I)/Fe(II)-based heterometallo-supramolecular polymer (**polyCuFe16**) was prepared by one-pot complexation of Cu(I), Fe(II), and an asymmetrical ligand (**L16**) (Fig. 6.3b) [3].

The coordination ability of ligands to metal ions is expressed by the complexation constant (K). In heterometallo-supramolecular polymers with two metal ion species, it is not easy to control the position of the two metal ions in the polymer chain, especially if the complexation constants of the two metal ions for the ligand do not differ considerably. One way of controlling precisely the position of the two metal ions in the polymer chain is to use an asymmetrical ligand (Fig. 6.3a).

An asymmetrical ligand with phenanthroline and terpyridine moieties (**L16**) was synthesized by the Suzuki–Miyaura cross-coupling reaction (Fig. 6.3c). A Miyaura-Ishiyama borylation reaction between 5-bromo-2,9-dimethyl-1,10-phenanthroline and bis(pinacolato)diboron (the molar ratio: 1:1) is carried out in the presence of PdCl$_2$(PPh$_3$)$_2$ as a catalyst for 6 h at 80 °C in DMSO to provide the corresponding boronic ester as a light yellow solid (80%). **L16** is obtained as a white solid (65%) using a Suzuki–Miyaura cross-coupling reaction of the boronic ester with 4′-(4-bromophenyl)-2,2′:6′,2″-terpyridine. **L16** is highly soluble in common organic solvents such as dichloromethane.

As the result of detailed titration experiments in the complexation of Cu(I) and Fe(II) with **L16**, it was revealed that Cu(I) and Fe(II) ions are preferentially complexed with the bidentate phenanthroline and tridentate terpyridine moieties of **L16**, respectively, based on the different (tetrahedral or octahedral) coordination geometries. These complexation constants for the formation of the 1:2 complexes are estimated to be quite high (more than 10^7). The opposite combinations of complexation do not occur in solution. Owing to the high complexation affinity and selectivity between the metal ions and the binding sites of the ligand, a

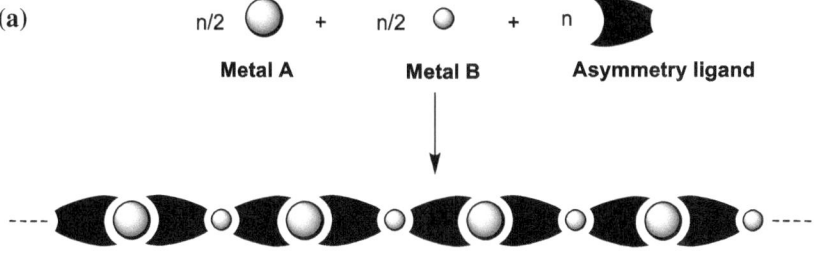

(a)

n/2 ⬤ Metal A + n/2 ◯ Metal B + n ◗ Asymmetry ligand

↓

Heterometallo-supramolecular polymer

(b)   n/2 Cu(I) + n/2 Fe(II) + n **L16**

↓

PolyCuFeL16

OTf⁻   2ClO₄⁻   n/2

(c)

Me, Br phenanthroline derivative + pinacol diboron (B–B), 1:1 ratio

PdCl₂(PPh₃)₂, KOAc, DMSO, 80 °C, 6 h

→ Me, boronic ester phenanthroline derivative

Br—phenyl-terpyridine

PdCl₂(PPh₃)₂, K₂CO₃, DMSO, 100 °C, 24 h

→ **L16**

Fig. 6.3 **a** Cu(I)/Fe(II)-based heterometallo-supramolecular polymer (**polyCuFeL16**) and **b** synthesis of **L16**

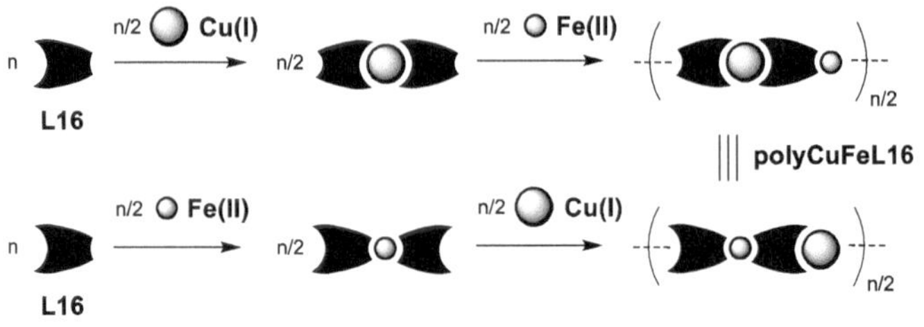

Fig. 6.4 A heterometallo-supramolecular polymer with Cu(I) and Fe(II) ions introduced alternately. **polyCuFeL16** is obtained regardless of the addition order of Cu(I) and Fe(II)

heterometallo-supramolecular polymer with Cu(I) and Fe(II) ions introduced alternately (**polyCuFeL16**) is obtained, regardless of the addition order of Cu(I) and Fe(II) (Fig. 6.4). Therefore, the preparation of **polyCuFeL16** is a one-pot method, prepared by the 1:0.5:0.5 complexation of an anhydrous dichloromethane solution of **L16** and anhydrous acetonitrile solutions of $Cu(CH_3CN)_4OTf$ and $Fe(ClO_4)_2 \cdot 6H_2O$ (Fig. 6.3b). The reaction mixture is stirred for 6 h under a nitrogen atmosphere at room temperature. **PolyCuFeL16** is obtained as a deep purple solid in 97% yield. It is highly soluble in acetonitrile. Molecular weight of the polymer couldn't be determined by GPC measurement because of the decomplexation in the GPC column.

6.4 Multi-Color Electrochromism

A thin film of **polyCuFeL16** is prepared by casting an acetonitrile solution. A cyclic voltammogram (CV) of the polymer film shows two redox waves based on Cu(I)/Cu(II) and Fe(II)/Fe(III) at 0.28 and 0.79 V versus Ag/Ag^+, respectively (Table 6.2). The redox potential of Cu(I)/Cu(II) in **polyCuFeL16** is much more negative than that in **polyCuL10** (0.41 V vs. Ag/Ag^+), but the difference in the redox potential of Fe(II)/Fe(III) is slight between **polyCuFeL16** and **polyFeL1** (0.82 V vs. Ag/Ag^+). In the metal complexes, a negative shift of the redox potential is often caused by an electron-donating effect of the ligand. Therefore, the large negative shift of the redox potential of Cu ions in **polyCuFeL16** is based on the intramolecular

Table 6.2 Redox potentials ($E_{1/2}$) of a Cu(I)/Fe(II)-based heterometallo-supramolecular polymer (**polyCuFeL16**) and the corresponding homometallo-supramolecular polymers (**polyCuL10** and **polyFeL1**)

	$E_{1/2}$ of Cu(I)/(II) (mV)	$E_{1/2}$ of Fe(II)/(III) (mV)
PolyCuFeL16	284	785
PolyCuL10	410	–
PolyFeL1	–	818

Working electrode: a glassy carbon electrode; reference electrode: Ag/Ag^+; counter electrode: Pt wire; electrolyte: 0.1 M TBAP/ethanol, scan rate: 50 mV/s

metal–metal interactions between the neighboring Cu and Fe ions: a weak electron-withdrawing effect of the neighboring Fe(II) complex moieties to Cu(I) ions in **poly-CuFeL16** resulted in the negative shift of the redox potential of Cu ions. On the other hand, the redox potential of Fe ions in **polyCuFeL16**, in which Fe(II) ions are oxidized to Fe(III), was similar to that in **polyFeL1**, because Cu(I) ions in **poly-CuFeL16** are already oxidized to Cu(II), which has a stronger electron-withdrawing effect than Cu(I).

A solid-state device with a **polyCuFeL16** film shows multi-color electrochromism (from purple to blue to colorless) when an applied voltage is changed from 0 to 2.2 V. In the absence of an applied voltage, the UV–vis spectrum shows two absorptions around 465 and 575 nm based on the MLCT absorptions of the Cu(I)-phenanthroline- and Fe(II)-terpyridine complex moieties, respectively. Only the absorption around 465 nm decreased at 1.4 V, because Cu(I) was oxidized to Cu(II) electrochemically and the MLCT absorption disappeared. The MLCT absorption around 575 nm also disappeared at 2.2 V, because Fe(II) is oxidized to Fe(III) at that voltage. The spectral change is reversible: the original spectrum is obtained by applying the opposite voltage (-2.4 V) to the device. For the oxidation of a **polyCuFeL16** film in the device at 2.4 V, the response times for the disappearance of the absorptions at 465 and 575 nm were 3.1 and 3.6 s, respectively. The electric conductivity of a **polyCuFeL16** film itself is too low to be determined by the general conductivity measurement.

References

1. Hossain MD, Sato T, Higuchi M (2013) Green color copper-based metallo-supramolecular polymer: synthesis, structure, and electrochromic properties. Chem Asian J 8:76–79. https://doi.org/10.1002/asia.201200668
2. Hossain MD, Higuchi M (2013) Synthesis of metallo-supramolecular polymers using 5,5′-linked bis(1,10-phenanthroline) ligands. Synthesis 45:753–758. https://doi.org/10.1055/s-0032-1316858
3. Hossain MD, Zhang J, Pandey RK, Sato T, Higuchi M (2014) A heterometallo-supramolecular polymer with Cu^{I} and Fe^{II} ions introduced alternately. Eur J Inorg Chem 2014:3763–3770. https://doi.org/10.1002/ejic.201402468

Chapter 7
Pt(II)-Based Metallo-Supramolecular Polymers

7.1 Synthesis

A Pt(II)-based metallo-supramolecular polymer with a head-to-tail polymer structure (**polyPtL17**) was prepared using a combination of an asymmetrical ligand and stepwise complexation (Fig. 7.1) [1].

An asymmetrical ligand **L17** is prepared as follows. 4′-(4-Bromophenyl)-2,2′:6′,2″-terpyridine (723 mg, 1.86 mmol) and pyridine-4-boronic acid (253 mg, 2.06 mmol) are dissolved in dry THF (12 mL) in a round-bottomed flask. After degassing, the solution is placed under an N_2 atmosphere. Potassium carbonate (767.3 mg, 5.56 mmol) is dissolved in the minimum volume of water in a separate round-bottomed flask and degassed similarly. The catalyst Pd(PPh$_3$)$_4$ (90 mg, 0.08 mmol) is added to the first round-bottomed flask, followed by the sodium carbonate solution. The solution is stirred at 85 °C and further portions of pyridine-4-boronic acid (50 mg, 0.41 mmol) and Pd(PPh$_3$)$_4$ (20 mg, 0.017 mmol) are added after 24 h. The solution is again stirred for another 48 h. The solvent is removed under reduced pressure and the crude residue is partitioned between dichloromethane and water. The organic layer is separated and washed with aqueous NaOH (0.1 M) (3 × 50 mL). After drying over sodium sulfate, the solvent is removed under vacuum to yield a brown solid (650 mg, 90%). Recrystallization from ethanol yields white crystals of the desired product.

The Pt(II)-based metallo-supramolecular polymer (**polyPtL17**) is prepared as follows. Dichloro-1,5-cyclooctadieneplatinum(II) (193 mg, 0.50 mmol) is treated with a solution of silver tetrafluoroborate (0.214 g, 1.1 mmol) in acetone/acetonitrile (5 mL, 4:1). The mixture is centrifuged to remove precipitated silver chloride, and the supernatant solution is added to a solution (methanol/water, 1:2) of **L17** (193.5 mg, 0.50 mmol) at 60 °C and stirred for 15 min in the dark. Then, the mixture is refluxed at elevated temperature overnight in the dark. The yellowish brown precipitate is filtered and washed with chloroform three times (50 mL × 3). The residue is dried to afford **polyPtL17** as a brown solid (320 mg, yield 84%). The head-to-tail structure of **polyPtL17** shown in Fig. 7.1b is supported by the peak assignment of the ^{1}H NMR

© National Institute for Materials Science, Japan 2019
M. Higuchi, *Metallo-Supramolecular Polymers*, NIMS Monographs,
https://doi.org/10.1007/978-4-431-56891-9_7

Fig. 7.1 **a** A head-to-tail metallo-supramolecular polymer formation. **b** Synthesis of Pt(II)-based metallo-supramolecular polymer (**polyPtL17**). **c** Synthesis of **L17**

and H-H COSY spectra as follows. ^{1}H NMR (300 MHz, DMSO-d$_6$, r.t.) δ ppm: 9.03 (s, proton-**g** in Fig. 7.1b), 8.92 (d, proton-**c**), 8.86 (d, proton-**f**) 8.73 (d, proton-**b**), 8.53 (dd, proton-**e**) 8.36 (d, proton-**h**) 8.13 (d, proton-**i**) 7.98 (dd, proton-**d**) 7.88 (d, proton-**a**). The low-intensity signals of the terminal pyridyl protons in the ^{1}H NMR spectrum of **polyPtL17** show that the degree of polymerization (DP) is 40 and that the molecular weight (MW) of the polymer is 23,240. Molecular weight was determined by viscometry–right-angle laser light scattering (RALLS); $M_w \sim 2.2 \times 10^4$ and $M_n \sim 2.0 \times 10^4$ (polydispersity index: 1.1). The value of M_n was similar to that estimated from the ^{1}H NMR spectrum.

7.2 Rectification

To investigate the electronic properties of **polyPtL17**, the metallo-supramolecular polymer was grafted onto functionalized Au electrodes to produce two films with their dipole moments aligned in different directions (Fig. 7.2). First, a monolayer of 4-(2,2′:6′,2″-terpyridin-4′-yl)benzenethiol (anchor A) or pyridine-4-thiol (anchor B) was attached to the Au electrode, and then **polyPtL17** was grafted to fabricate two films. In **Film 1**, the dipole moment is oriented away from the Au electrode, whereas, in **Film 2**, the dipole moment is oriented towards the Au electrode. To confirm the surface derivatization of the Au electrode by **polyPtL17**, the films were examined by UV–vis and Raman spectroscopies and CV.

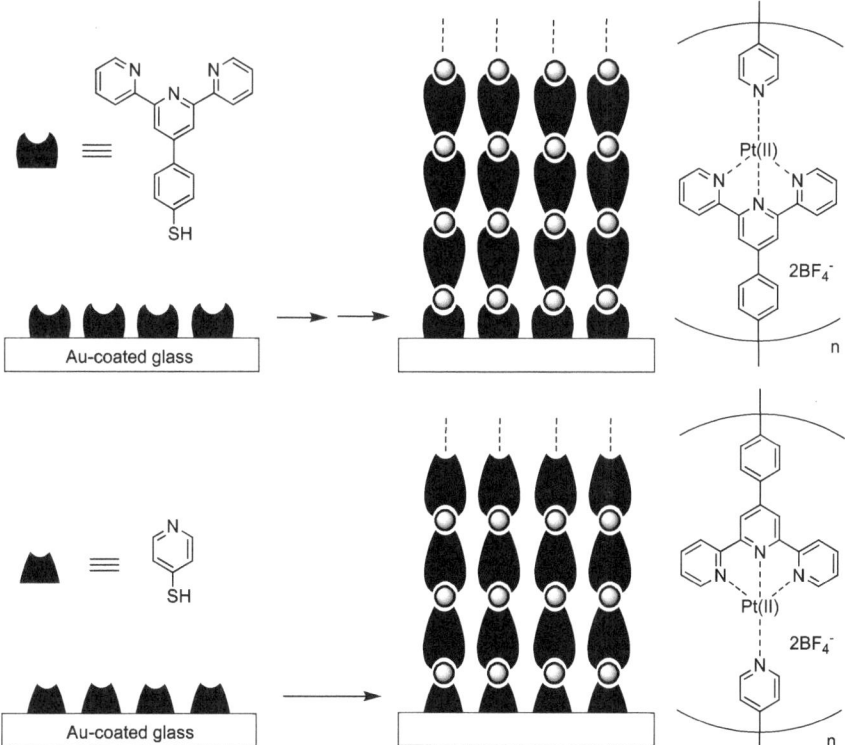

Fig. 7.2 Polymer films formed with controlled dipole moment by anchoring **polyPtL17**

Film 1: A 10-nm-thick Au (111)-coated glass was rinsed in ethanol and immersed in 1 mM ethanol solution of anchor A for 24 h. The Au substrate was taken out and rinsed with excess ethanol and dried with an N_2 stream. A monolayer of anchor A was formed over the Au-coated glass. Then, it was again immersed in an ethanol solution of 1 mM Pt(COD)Cl$_2$ at 60 °C for 30 min. The Au-coated glass was taken out and again rinsed with excess ethanol. Finally, it was immersed in an ethanol solution of 1 mM **polyPtL17** under reflux condition for 24 h. The Au-coated glass was taken out and rinsed with excess ethanol and finally dried with an N_2 stream to produce **Film 1**.

Film 2: A 10-nm-thick Au (111)-coated glass was rinsed in ethanol and immersed in a 1 mM ethanol solution of anchor B for 24 h. The Au substrate was taken out and rinsed with excess ethanol and dried with an N_2 stream. A monolayer of anchor B was formed over the Au-coated glass. Then, it was again immersed in an ethanol solution of 1 mM **polyPtL17** and refluxed for 24 h. The Au-coated glass was taken out and again rinsed with excess ethanol. It was finally dried with an N_2 stream to produce **Film 2**.

To complete the fabrication of the device, a second 10-nm-thick Au strip was deposited orthogonally to the first Au electrode to form a sandwich of the polymer between the two Au electrodes in each film.

The electronic properties of the two films were determined by current–voltage ($I–V$) measurements. The $I−V$ characteristics of the two films showed an asymmetric charge-transport behavior resulting from intrinsic polarization in the oriented sandwiched polymer chain as a result of its inherent dipole. From quantum mechanical calculations of the dipole moment and frontier molecular orbitals, it was determined that the average dipole moment for each monomer unit in **polyPtL16** was about 5.8 D. The nonsymmetric charge transport results of the diode films can qualitatively be explained by considering the effect of bond dipole on the alignment of the Fermi energy (E_F) levels of the electrodes. The bond dipole can affect the surface dipole of the metal electrode, and a local charge rearrangement has been demonstrated to be induced by an anchored polymer binding a metal electrode. When a polymer chain is connected to electrodes, the surface dipole layer induces a shift in the vacuum level (V_L) at the surface of the gold electrodes, which alters the alignment of the Fermi level (E_F) at the interface. However, the most interesting result is that the charge-transport behavior in the two films was exactly opposite. **Film 1** was more conductive under a positive bias, with an average rectification ratio RR $= I$(+4 V)/I(-4 V) $\approx$ 20, whereas **Film 2** was more conducting under a negative bias, with an average rectification ratio RR $= I$(-4 V)/I(+4 V) $\approx$ 18.

In **Film 1**, under a positive bias, the direction of the current flow between the two electrodes was parallel to the internal dipole, as the dipole moment was oriented away from the bottom Au electrode. Because both directions were aligned, **Film 1** allowed a high current to pass in the positive bias region. However, under negative bias, the direction of the current flow between the two electrodes was opposite to that of the internal dipole, and consequently, **Film 1** allowed a very low current to pass under negative bias. As **Film 2** had its dipole direction oriented towards the bottom Au electrode (i.e., the direction opposite to Film 1), it showed $I-V$ properties exactly opposite to those of **Film 1**. **Film 2** allowed a high current to pass only under negative bias, as the direction of the current flow between two electrodes was then parallel to the direction of the internal dipole. This type of analysis has been shown mechanistically in previously reported multilayer unimolecular rectifier device systems. Although the exact position of E_F in a molecular system is very difficult to determine, we can assume that the Fermi levels of Au electrodes would align at the middle of the HOMO and LUMO energy gap of the molecule under a zero bias and no internal dipolar field. For **Film 1**, as the direction of dipole moment was away from the Au electrode, it pushed down the E_F of the Au electrode when the device was connected to an outside circuit. This shift could lead to electron tunneling from the HOMO of the diode of **Film 1**, as the E_F of the Au electrode was closer to the HOMO of the molecular diode film. On the other hand, as **Film 2** had a dipole moment direction opposite to that of Film 1, it could increase the E_F of the Au electrode to produce an effective electron tunneling from the LUMO of the diode **Film 2** as the E_F of the Au electrode was closer to the LUMO of the molecular diode film. Therefore, owing to the contrary dipole moment direction of the two films, the opposite rectifying behaviors of **Film 1** and **Film 2** were obtained.

Reference

1. Chakraborty C, Pandey RK, Hossain MD, Futera Z, Moriyama S, Higuchi M (2015) Platinum(II)-based metallo-supramolecular polymer with controlled unidirectional dipoles for tunable rectification. ACS Appl Mater Interfaces 7:19034–19042. https://doi.org/10.1021/acsami.5b03434

Chapter 8
Ni(II)-Based Metallo-Supramolecular Polymers

8.1 Synthesis

To introduce Ni(II) ions with a square planar coordination geometry to a metallo-supramolecular backbone, phenanthroline with bidentate coordination ability is a proper coordination site in the ditopic ligands. Actually, the 1:1 complexation of Ni(ClO$_4$)$_2$·6H$_2$O and **L14** was confirmed by a UV–vis spectrophotometric titration. The ligand-centered absorption (λ_{max} = 274 nm) of **L14** is shifted to 282 nm during the addition of up to 1.0 equiv. of the Ni(II) salt due to complexation, but the spectral change became saturated (no longer changed) with the addition of more than 1.1 equiv. The titration plots between the absorbance and the molar ratio of [Ni(II)]/[**L14**] clearly show the 1:1 complexation of the Ni salt with **L14**.

Ni(II)-based metallo-supramolecular polymers (**polyNiL12-15**) are synthesized by the 1:1 complexation of Ni(II) ions with **L12-15** (Fig. 8.1) [1]. The detailed synthetic procedure is as follows. The ligand (**L12-15**) (0.026 mmol) is dissolved in 5 mL of CH$_2$Cl$_2$ under a nitrogen atmosphere. Ni(ClO$_4$)$_2$·6H$_2$O (0.026 mmol) dissolved in 5 mL of CH$_3$CN is added dropwise over 30 min to the ligand solution.

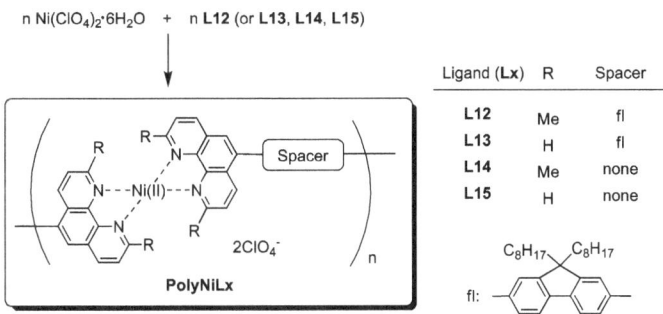

Fig. 8.1 Synthesis of Ni(II)-based metallo-supramolecular polymers (**polyNiL12-15**) by the 1:1 complexation of Ni(ClO$_4$)$_2$ H$_2$O with **L12-15**

© National Institute for Materials Science, Japan 2019
M. Higuchi, *Metallo-Supramolecular Polymers*, NIMS Monographs,
https://doi.org/10.1007/978-4-431-56891-9_8

The mixture is stirred for 1 h at room temperature. The solvent is slowly evaporated from the pink solution by passing a stream of nitrogen over the reaction mixture. During that procedure, a precipitate forms. It is rinsed thoroughly with 20 mL of Et$_2$O, filtered, then dried in vacuo at room temperature to afford the desired polymer **polyNiL12-15** (**polyNiL12**: pink solid; yield: 92%; M_w: 1.24 × 10^5 Da; **polyNiL13**: pink solid; yield: 95%; M_w: 1.05 × 10^5 Da; **polyNiL14**: pink solid; yield: 90%; M_w: 1.28 × 10^5 Da; **polyNiL15**: pink solid; yield: 93%; M_w: 1.03 × 10^5 Da). The molecular weights (M_w) were determined by SEC-viscometry–RALLS.

8.2 Ionic Conductivity

Ionically conductive materials are widely used as electrolytes in fuel cells and secondary batteries. The formation of ion-conduction channels in the materials is required to achieve high ionic conductivity. For instance, hydrophilic sulfonic acid groups in a hydrophobic Nafion film form inverse-cylindrical micellar structures as ion channels. A balance of hydrophilicity and hydrophobicity in the materials is required to control the ionic conductivity. Ionic conductivity in Ni(II)-based metallo-supramolecular polymers (**polyNiL12-14**) was observed and the effect of the ligand modification was investigated.

The Nyquist plots for the polymer films were obtained by an ac impedance measurement at room temperature and 98 %RH (relative humidity). **PolyNiL14** without a hydrophobic spacer showed an ionic conductivity about 500 times higher than that of **polyNiL12** with a hydrophobic spacer. **PolyNiL13** without methyl groups in the phenanthroline moieties showed an ionic conductivity about 20 times higher than that of **polyNiL12** with methyl groups (Table 8.1). The introduction of hydrophobic units to the ligand markedly decreased the ionic conductivity of the polymer. These results indicate that the formation of ion channels in the polymer film at high humidity was partially prevented by the bulky hydrophobic moieties. The electrical conductivity

Table 8.1 Ionic conductivity and activation energy of Ni(II)-based metallo-supramolecular polymers (**polyNiL12-14**)

	Ion conductivity at 98 %RH at r.t. (mS/cm)	Activation energy (E_a) (eV)
PolyNiL12	1.44 × 10^{-3}	–
PolyNiL13	3.2 × 10^{-2}	–
PolyNiL14	0.75	0.43

The polymer films were prepared by casting the MeCN/EtOH (1:1) solution on conductivity-measuring electrodes (CMEs). However, a uniform film of **polyNiL15** could not be obtained owing to the low solubility in organic solvents. The film thickness was determined using ellipsometry to be 24.2, 21.5 and 16.0 nm for **polyNiL12-15** films, respectively, and the average roughness of the **polyNiL14** film was about 4.5 nm. The activation energy (E_a) was determined on the basis of the ionic conductivity measured at different temperatures at 98 %RH (applied voltage: 1.0 V). E_a was calculated from the slope of the Arrhenius plot

of the polymers was very low: only a negligible current was detected by applying a voltage up to 6 V under vacuum-dried conditions. The ionic conductivity of the polymers depended strongly on the humidity: the log of the ionic conductivity as a function of relative humidity (%RH) in a **polyNiL14** film showed an almost linear dependence with an average slope of 0.046. A range of slopes around 0.07 ± 0.03 has been reported for ionically conductive organic polymers. The activation energy (E_a) for ionic conduction in a **polyNiL14** film was investigated by measuring the dc current response as a function of temperature. E_a was determined to be 0.43 eV from the Arrhenius plot. The low activation energy indicates proton conduction based on the Grotthuss mechanism at high humidity.

8.3 Preparation of Polymers with Different Counter Anions

To investigate the effect of the counter anion on ionic conductivity of metallo-supramolecular polymers, Ni(II)-based metallo-supramolecular polymers with different counter anions (chloride (Cl^-), nitrate (NO_3^-), acetate (CH_3COO^-), and acetylacetonate ($CH_3C(=O)CH=C(O^-)CH_3$)) (**polyNiL14-Cl, -NO$_3$, -ac,** and **-acac,** respectively) were prepared by the 1:1 complexation of the Ni(II) salts and **L14** (Fig. 8.2) [2]. The synthetic procedure of **polyNiL14-Cl** is as follows. An acetonitrile solution (5 mL) of $NiCl_2$ (0.026 mmol) is added dropwise over 30 min to a CH_2Cl_2 solution (5 mL) of **L14** (0.026 mmol) under nitrogen atmosphere. The reaction mixture is stirred for 1 h at room temperature. Then, the solvent is slowly evaporated from the pink solution by passing a stream of nitrogen over the reaction mixture. **PolyNiL14-Cl** is obtained as precipitate during the evaporation and is rinsed with 20 mL of Et_2O, filtered, then dried in vacuo at room temperature. **PolyNiL14-Cl, -NO$_3$, -ac,** and **-acac** are light yellow or yellow solids, and the yields are 92–98%. The molecular weights (M_w) were determined by SEC-viscometry-RALLS (solvent: acetonitrile; polymer concentration: 1.0 mg/mL; injection volume:

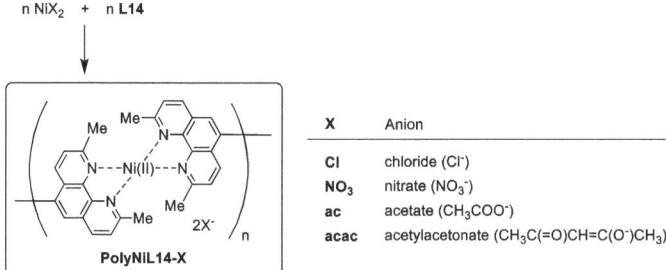

Fig. 8.2 Synthesis of Ni(II)-based metallo-supramolecular polymers with different counter anions (**polyNiL14-X**)

20 μL; standard: polyethylene oxide-PEO-22 K) to be 1.23×10^5 Da (**polyNiL14-Cl**), 1.18×10^5 Da (**polyNiL14-NO$_3$**), 1.30×10^5 Da (**polyNiL14-ac**), and 1.56×10^5 Da (**polyNiL14-acac**).

The 1:1 complexation of NiCl$_2$·6H$_2$O and **L14** was also confirmed by UV–vis spectrophotometric titration experiments. A ligand-centered absorption ($\lambda_{max} = 273.4$ nm) in **L14** was shifted to longer wavelength ($\lambda_{max} = 280$ nm) by the addition of the Ni(II) salt. This shift occurs during the addition of 0.0–1.0 equiv. of Ni(II) salt, and no spectral change was observed by the further addition of the metal salt. When the absorbance at 288.4 nm is plotted as a function of the molar ratio ([Ni(II)]/[**L14**]), the absorbance increased in proportional to the molar ratio, and then was saturated at a molar ratio of 1.0. This result suggests that the 1:1 complexation of Ni(II) ions and the ligand occurred in solution.

8.4 Effect of Counter Anions on Ionic Conduction

Ionic conductivity is calculated from the diameter of the semicircle at the high-frequency region of Nyquist plots (Table 8.2). The Nyquist plots of the polymer films are obtained from the ac impedance measurements at 25 °C and 98 %RH. The polymer with "harder" anions shows higher ionic conductivity: the order of the hardness of the anions (chloride > nitrate > acetate > acetylacetonate) is in good agreement with the order of the ionic conductivity of the polymers (**polyNiL14-Cl >** **polyNiL14-NO$_3$** > **polyNiL14-ac** > **polyNiL14-acac**). The ionic conductivity of the polymer films greatly depends on the humidity and increased by about four orders of magnitude when the humidity was increased from 30 to 98 %RH (Table 8.3). Their activation energies for the ionic conductivity were determined using Arrhenius plots (log of current as a function of 1/T) (Table 8.2).

An activation energy lower than 0.4 eV suggests the Grotthuss mechanism, in which protons pass along hydrogen bonds. These results indicate that the high ionic

Table 8.2 Ionic conductivities and activation energies of Ni(II)-based metallo-supramolecular polymers (**polyNiL14-Cl**, **-NO$_3$**, **-ac**, and **-acac**)

	Ion conductivity at 98 %RH at r.t. (S/cm)	Activation energy (E_a) (eV)
PolyNiL14-Cl	5.0×10^{-2}	0.21
PolyNiL14-NO$_3$	2.4×10^{-3}	0.31
PolyNiL14-ac	1.0×10^{-3}	0.32
PolyNiL14-acac	0.6×10^{-3}	0.80

The polymer films were prepared on CME with an electrode gap of 10 μm by drop-casting the polymer solution [solvent: acetonitrile/ethanol (1:1)]. The film thicknesses of **polyNiL14-Cl**, **polyNiL14-NO$_3$**, **polyNiL14-ac**, and **polyNiL14-acac** were 30.0, 36.1, 68.8, and 67.1 nm, respectively, as determined by ellipsometry. The activation energy (E_a) was determined on the basis of the ionic conductivity measured at 98 %RH and different temperatures (20–80 °C). E_a was calculated from the slope of the Arrhenius plot

Table 8.3 Ionic conductivities of Ni(II)-based metallo-supramolecular polymers (**polyNiL14-Cl**, **-NO₃**, **-ac**, and **-acac**) at different %RHs

%RH	Ionic conductivity (S/cm)			
	PolyNiL14-Cl	**PolyNiL14-NO$_3$**	**PolyNiL14-ac**	**PolyNiL14-acac**
30	5.0×10^{-6}	5.4×10^{-7}	4.5×10^{-8}	4.0×10^{-8}
40	3.8×10^{-5}	3.1×10^{-6}	4.5×10^{-7}	5.3×10^{-7}
50	1.2×10^{-4}	9.1×10^{-6}	2.6×10^{-6}	3.7×10^{-6}
60	2.8×10^{-4}	3.3×10^{-5}	1.2×10^{-5}	1.4×10^{-5}
70	6.5×10^{-4}	9.5×10^{-5}	3.7×10^{-5}	3.9×10^{-5}
80	1.6×10^{-3}	1.9×10^{-4}	1.6×10^{-4}	1.0×10^{-4}
90	4.6×10^{-3}	1.0×10^{-3}	5.2×10^{-4}	2.2×10^{-4}
100	5.0×10^{-2}	2.4×10^{-3}	1.0×10^{-3}	6.1×10^{-4}

conductivity in **polyNiL14-Cl**, **-NO₃**, and **-ac** films at high humidity is achieved by efficient proton transfer through the ion channels formed by water molecules, which are assembled along the polymer chains with the help of the positive charge of Ni(II) ions and/or the negative charge of the counter anions. An activation energy higher than 0.4 eV in a **polyNiL14-acac** film, on the other hand, indicates proton transfer aided by a moving vehicle (additional ions or molecules), probably because the bulky anions prevent proton channel formation through the polymer chains.

8.5 Real-Time Humidity Sensors

Capacitive, thermal, and resistive methods for moisture detection in air are found in the literature. In addition, Ni(II)-based metallo-supramolecular polymers serve as an efficient and rapid dynamic sensing device for humidity change because of the high responsiveness of ionic conductivity to humidity. A **polyNiL14-Cl** film was prepared on an interdigitated electrode (Fig. 8.3) and the current was measured in the humidity range from 25 to 98 %RH. When the humidity was gradually swept from 25 to 90 %RH, an increase in the conductance (1/R) was observed. In the humidity

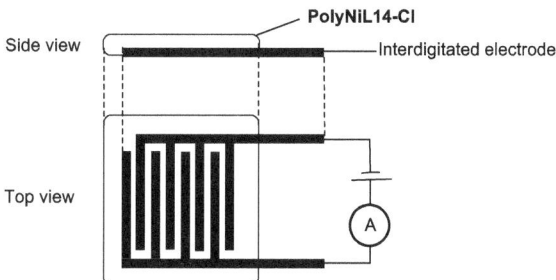

Fig. 8.3 A real-time humidity sensor with **polyNiL14-Cl** (an interdigitated electrode composed of 65 pairs of Pt wires printed on a quartz substrate with 10 μm gaps between each Pt electrode pair. The bias voltage was fixed at 1.0 V)

range higher than 95 %RH, the conductance became almost constant. The dynamic sensitivity to humidity was also confirmed by another experiment: when the ionic conductance was measured during the stepwise increase in humidity from 25 to 98 %RH (30 min duration) and vice versa, a stepwise change in the ionic conductance was observed corresponding to the relative humidity change. The gradual decrease in current at each humidity level was probably due to the ionic polarization of the polymer film caused by application of voltage over a long time. The ion accumulation near the electrodes increased the potential gap between the electrodes and decreased the current value. The polymer film itself showed reproducible and stable behavior even after the experiment was continuously carried out for 5–6 h.

References

1. Pandey RK, Hossain MD, Moriyama S, Higuchi M (2013) Ionic conductivity of Ni(II)-based metallo-supramolecular polymers: effects of ligand modification. J Mater Chem A 1:9016–9018. https://doi.org/10.1039/c3ta12080e
2. Pandey RK, Hossain MD, Moriyama S, Higuchi M (2014) Real-time humidity-sensing properties of ionically conductive Ni(II)-based metallo-supramolecular polymers. J Mater Chem A 2:7754–7758. https://doi.org/10.1039/c4ta00884g

Chapter 9
Mo(VI)-Based Metallo-Supramolecular Polymer

9.1 Synthesis

Chemical reactions with polymers (polymer reactions) are often performed to improve or modify the properties of the original polymer. It is considered that the polymer reactions can be adapted to metallo-supramolecular polymers (Fig. 9.1a), but one must be careful about decomplexation in the polymer chains during the reaction. A Mo(VI)-based metallo-supramolecular polymer (**polyMoL12**) was prepared by the 1:1 complexation of bis(phenanthroline) (**L12**) with $MoO_2(acac)_2$ (Fig. 9.1b) [1]. Since the resultant polymer had four methyl groups for each ligand, the oxidation of the methyl groups was investigated.

The synthetic procedure for **polyMoL12** is as follows. **L12** (0.03 mmol) is dissolved in CH_2Cl_2 (5 mL) under a nitrogen atmosphere. $MoO_2(acac)_2$ (0.03 mmol) dissolved in CH_3CN (5 mL) is added to the ligand solution dropwise over a period of 30 min. The mixture is stirred for 12 h at 40 °C. Then, the solvent is slowly evaporated from the pink solution by passing a stream of nitrogen over the reaction mixture. The precipitate is rinsed thoroughly with Et_2O (20 mL), filtered, then dried in vacuo at room temperature to afford **polyMoL12** in 90% yield. The polymer with carboxylic acid groups (**polyMoL12-C**) is obtained by the oxidation of **polyMoL12** as follows. A two-necked flask fitted with a stirrer bar and an oxygen gas inlet and outlet is charged with **polyMoL12** (0.0123 mmol) and anhydrous DMF (2 mL). A solution of t-BuOK (0.125 mmol) in DMF (2 mL) is added dropwise and the yellow mixture is stirred at r.t. for 1 h. Oxygen is then purged into the reaction mixture for 6 h. The solvent is slowly evaporated from the orange solution by passing a stream of nitrogen over the reaction mixture. The precipitate is rinsed thoroughly with H_2O (20 mL), filtered, then dried in vacuo at room temperature to afford orange-colored **polyMoL12-C** in 85% yield. The chemical transformation from **polyMoL12** to **polyMoL12-C** is supported by the ^1H NMR spectra. A new broad peak attributed to acid protons (-COOH) appeared at δ 11.97 ppm in **polyMoL12-C**. The degree of oxidation was calculated to be 84% by comparing the relative integration value of this broad peak with a peak based on an aromatic

© National Institute for Materials Science, Japan 2019

M. Higuchi, *Metallo-Supramolecular Polymers*, NIMS Monographs,

https://doi.org/10.1007/978-4-431-56891-9_9

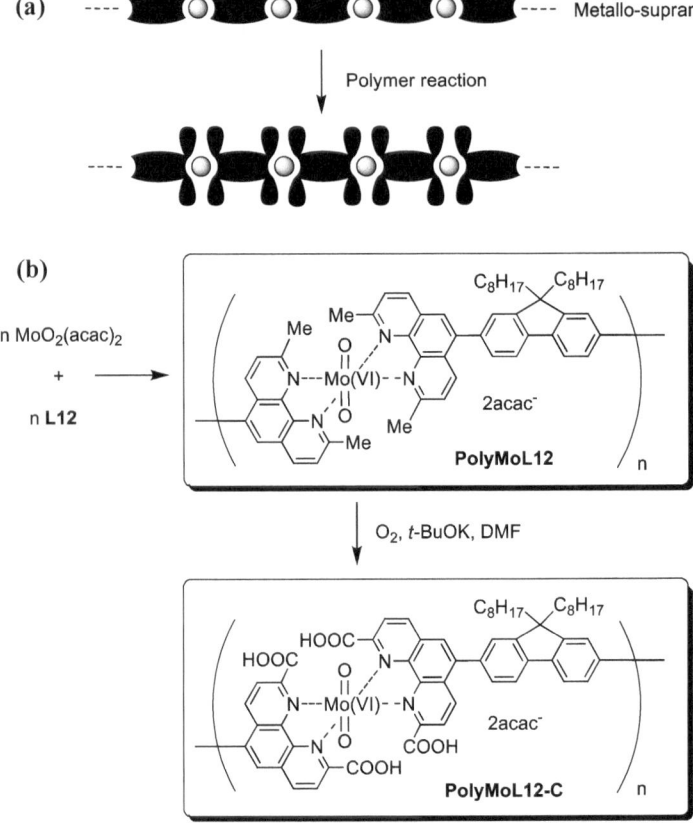

Fig. 9.1 a Modification of a metallo-supramolecular polymer structure by a polymer reaction.
b Synthesis of Mo(VI)-based metallo-supramolecular polymer (**polyMoL12**) and its oxidized
derivative (**polyMoL12-C**)

proton. Actually, the degree of oxidation was judged to be more than 84% because
acid protons often appear as smaller peaks than the expected size owing to the dis-
sociation of the acids. Molecular weight of **polyMoL12** is almost unchanged by the
oxidation. Weight-average molecular weights of **polyMoL12** and **polyMoL12-C** are
determined by SEC-viscometry–RALLS system to be 1.13×10^5 Da and 1.02×10^5
Da, respectively, using polyethylene oxide-PEO-22 K as standard. The molecular
weights are obtained by automatic program calculation taking account of viscosity
and RALLS factor into consideration.

9.2   Proton Conduction

The Nyquist plots of **polyMoL12** and **polyMoL12-C** were obtained from ac impedance measurements. The ionic conductivities of the polymers were calculated from the x-axis intercept of the semicircle at higher frequencies (Table 9.1). The introduction of carboxylic acids to the ligand enhanced the ionic conductivity of the polymer more than tenfold. **PolyMoL12** exhibited quite low conductivity at low humidity. More than five orders of growth in the conductivity of **polyMoL12** was observed with increasing humidity. Both polymers contain acetylacetonate (acac) as counter anions. The mobility of acac in the polymer films should be very low owing to both its bulky structure and ionic interactions with the metal center Actually, the very poor contribution of acac to ionic conduction in metallo-supramolecular polymer films has been established in our previous paper, although acac can provide a pathway for proton transfer in polymer films due to its keto-enol interconversion process (tautomerism). Considering the definite increase in conductivity with humidity, the conductivity is solely attributed to the ionic species originating due to humidity i.e., water, which could either be protons or hydroxyl ions. Water molecules can assemble themselves in certain arrangements because of the interaction with the hydrophilic central metal ions. The orderly arrangement of water molecules is known in the literature for proton-conducting materials. To further probe this phenomenon, an FTIR study of the polymers was carried out at two different humidities. FTIR spectra have earlier been used to analyze the protonation sites in proton-conducting materials. The hydrogen-bonded OH stretch, H-bonding vibration, and free OH have been used to define the existence of the H-bonded network and/or water clusters in materials having high proton conductance. It is conceivable that the protons combine with other water molecules to form hydronium ions, or Zundel or Eigen cations under certain conditions. In the FTIR spectra of **polyMoL12** and **polyMoL12-C** at high humidity, the symmetric and asymmetric free OH stretches are clearly seen in the

Table 9.1 Ionic conductivity and activation energy of Mo(VI)-based metallo-supramolecular polymers (**polyMoL12** and **polyMoL12-C**)

	Ion conductivity at 98 %RH at 25 °C (mS/cm)	Activation energy (E_a) (eV)
PolyMoL12	3.2×10^{-3}	1.80
PolyMoL12-C	8.5×10^{-2}	0.54

The powdered polymer samples were pressed in the form of pellets, and these pellets were inserted between the two electrodes of a sample holder, SH-2Z, from Toyo Industries, Tokyo, to enable precise conductivity measurements. The sample holder was attached to a sensitive micrometer to measure pellet thickness. The entire setup was then placed inside a humidity-temperature controller, SH-221 of Espec Japan. Prior to each measurement, the pellet was held for at least 3–4 h at the desired temperature and humidity. Subsequently, the conductivity was measured using a Solartron 1260 impedance gain/phase analyzer coupled with a Solartron 1296 dielectric interface. A frequency range of 1 Hz to 10 MHz was used to determine the resistivity of the polymer pellet. The activation energy (E_a) was determined on the basis of the ionic conductivity measured at 95 %RH and different temperatures. E_a was calculated from the slope of the Arrhenius plot

region beyond $3500\ cm^{-1}$. The spectra were fit using Gaussian peak fitting in the OH stretching region. A significant intensity increment in the OH stretching region for the hydrated **polyMoL12** indicates the integration of –OH into the material structure, which could be crucial for proton conduction. These peaks signify the presence of bigger clusters of water in the form of $H^+(H_2O)_n$ in the polymer, especially when the polymer is hydrated. The FTIR spectrum of **polyMoL12-C** also shows similar peaks in addition to the peak for the C=O stretch. In the case of **polyMoL12-C**, these peaks can be seen even at lower humidities, which is reasonable owing to the fact that the presence of hydrophilic carboxylic acid functional groups in **polyMoL12-C** helps to retain water molecules in the polymer, even at lower humidities. The OH stretching frequencies in **polyMoL12-C** are slightly shifted towards blue in comparison with those in **polyMoL12**. It is known that a blue shift in the frequency of OH stretching peaks is usually associated with an increase in n in $H^+(H_2O)_n$. This is crucial evidence signifying that **polyMoL12-C** has a larger n and consequently a higher proton conductivity. It also sheds some light on the occurrence of diffusional impedance, as seen in the Nyquist plot for **polyMoL12-C**. We also observed a reversible transition in the appearance of OH stretching peaks in the presence and absence of humidity.

Reference

1. Pandey RK, Hossain MD, Chakraborty C, Moriyama S, Higuchi M (2015) Proton conduction in Mo(VI)-based metallo-supramolecular polymers. Chem Commun 51:11012–11014. https://doi.org/10.1039/c5cc03634h

Chapter 10
Cd(II)-Based Metallo-Supramolecular Polymer

10.1 Synthesis

Cd(II)-based metallo-supramolecular polymer (**polyCdL12**) (Fig. 10.1) is prepared as follows [1]. **L12** (0.03 mmol) is dissolved in CH_2Cl_2 (5 mL) under a nitrogen atmosphere. $Cd(ClO_4)_2 \cdot 6H_2O$ (0.03 mmol) dissolved in acetonitrile (5 mL) is added to the ligand solution dropwise over 30 min. The mixture is stirred for 1 h at room temperature. The solvent is slowly evaporated from the yellow solution by passing a stream

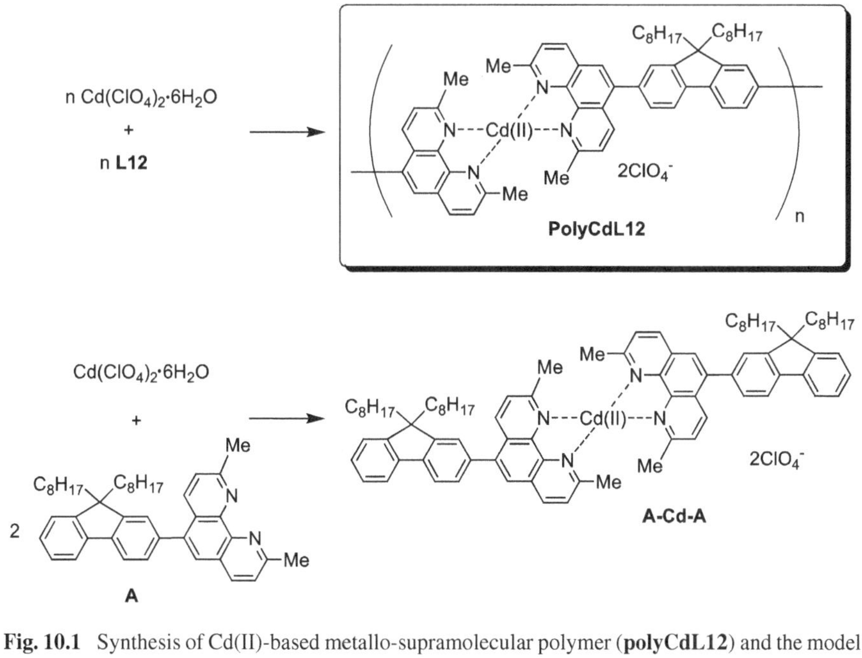

Fig. 10.1 Synthesis of Cd(II)-based metallo-supramolecular polymer (**polyCdL12**) and the model complex (**A-Cd-A**)

© National Institute for Materials Science, Japan 2019
M. Higuchi, *Metallo-Supramolecular Polymers*, NIMS Monographs,
https://doi.org/10.1007/978-4-431-56891-9_10

of nitrogen over the reaction mixture. During that procedure, a precipitate forms and is rinsed thoroughly with 20 mL of Et_2O, filtered, and dried in vacuo at room temperature to afford **polyCdL12** (yield: 95%) as a deep yellow solid. Molecular weight: 3.7×10^5 Da, which was determined by SEC-viscometry–RALLS (size exclusion chromatography-viscometry-right angle light scattering solvent) system consisting of a pump, solvent degasser, liquid chromatograph, refractive index detector, column oven, viscotek 270 dual detector. The eluent was acetonitrile at a flow speed of 1 mL/min. The column temperature was 30 °C. The synthesized polymers (c = 1.0 mg/mL) show weight-average molecular weight using polyethylene oxide-PEO-22 K as standard, when 20 μL of acetonitrile solution was injected. The molecular weight was obtained by automatic program calculation taking account of viscosity and RALLS factor into consideration.

As the reference for **polyCdL12**, the monometal complex (**A-Cd-A**) was also synthesized as follows. **A** (0.08 mmol) is dissolved in acetonitrile (5 mL) under a nitrogen atmosphere. $Cd(ClO_4)_2 \cdot 6H_2O$ (0.04 mmol) dissolved in acetonitrile (5 mL) is added to the ligand solution dropwise over 30 min. The mixture is stirred for 1 h at room temperature. The solvent is slowly evaporated from the green solution by passing a stream of nitrogen over the reaction mixture. During that procedure, a green precipitate forms and is rinsed thoroughly with 20 mL of Et_2O, filtered, and dried in vacuo at room temperature to afford **A-Cd-A** (yield: 91%).

10.2 Nano-Molar Detection of Cd(II) Ions

The binding constants $(\log K)$ for **polyCdL12** and **A-Cd-A** were calculated to be 5.97 and 2.68, respectively, using the modified Benesi–Hildebrand equation (10.1).

$$\frac{(I_{max} - I_0)}{(I_x - I_0)} = 1 + \left(\frac{1}{K}\right)\left(\frac{1}{[M]^n}\right) \tag{10.1}$$

Here, I_{max}, I_x, and I_0 are maximum absorbance intensity (saturated) of the ligand in the presence of Cd(II), the absorbance intensity of the ligand in the presence of Cd(II) at an intermediate concentration, and the free ligand absorbance, respectively. When K is the binding constant, $[M]$ represents the concentration of Cd(II), and n is the number of metals bound per ligand (for **polyCdL12**, n = 1; for **A-Cd-A**, n = 0.5). The highly sensitive and selective detection of toxic metal ions such as Cd(II) is required in clinical care and diagnosis. As for Cd(II) sensing, the reported detection limit is 25 nM. We found that **L12** could detect Cd(II) at the level of 8 nM by forming the metallo-supramolecular polymer, which suggests the possibility of practical applications in the area of Cd(II) ion detection. The luminescence of **L12** originates from a π-π* transition of the fluorine unit (Table 10.1). The emission wavelength is red-shifted by 1:1 complexation with Cd(II) ions. **PolyCdL12** shows a green emission. In general, metal coordination increases the electron-accepting ability of the ligand, and decreases and stabilizes the electron transition energy of

Table 10.1 Photophysical properties of the ligands and polymers

	λ_{max} (nm) (ε, M^{-1}cm^{-1})	λ_{max} (ex)	λ_{max} (em)	ϕ_{em}
L12	326 (52,600), 280 (71,600)	326	421	0.52
PolyCdL12	330 (34,200), 280.5 (64,100)	330	492	0.34
A	311 (22,700), 271.5 (37,400)	311	397	0.18
A-Cd-A	338 (11,900), 272 (44,900)	338	489	0.09

Ligands (**L12** and **A**): 1×10^{-5} M in CH$_2$Cl$_2$; Cd complexes (**polyCdL12** and **A-Cd-A**): 1×10^{-5} M in CH$_3$CN

the intraligand charge transfer. In the case of **polyCdL12**, the structural rigidity of the metallo-supramolecular polymers seems to be higher than that of the ligand, are responsible for a decrease in energy loss, and make them suitable to emit fluorescence. **A-Cd-A** exhibits a weaker emission (quantum yield: 0.09) than **polyCdL12** (quantum yield: 0.34) (Table 10.1). The enhancement of the luminescence of **polyCdL12** with respect to **A-Cd-A** is attributed to the strong chelation by the ligands to the Cd in the polymer, which effectively increases the rigidity of the ligand and reduces the loss of energy. Interestingly, the fluorescence of **L12** is totally quenched by 1:1 complexation with other four-coordinate transition metal ions such as Ag(I), Cu(I), Cu(II), Ni(II), and Pd(II). The quenching can be caused by processes such as magnetic perturbation, redox activity, electronic energy transfer, and donation of lone pairs of electrons.

Reference

1. Hossain MD, Pandey RK, Rana U, Higuchi M (2015) Nano molar detection of Cd(II) ions by luminescent metallo-supramolecular polymer formation. J Mater Chem C 3:12186–12191. https://doi.org/10.1039/c5tc02734a

Chapter 11
Eu(III)-Based Metallo-Supramolecular Polymer

11.1 Synthesis

A Eu(III)-based metallo-supramolecular polymer (**polyEuL18**) is prepared by the 1:1 complexation of Eu(III) ions with **L18** (Fig. 11.1) [1].

Unlike transition metal ions, lanthanide ions such as Eu(III) ions have 8–10 coordination sites. To introduce such metal ions into a metallo-supramolecular polymer backbone, synthesis of a new ditopic ligand (**L18**) is necessary.

Compound **B** (diethyl-4'-(4-bromophenyl)-2,2':6',2''-terpyridine-6,6''-dicarboxylate) is synthesized as follows. A mixture of 4'-(4-bromophenyl)-2,2':6',2''-terpyridine-6,6''-dicarbonitrile (8.77 g, 20.0 mmol), conc. H_2SO_4 (90 mL), acetic acid (90 mL), and H_2O (20 mL) is stirred at 100 °C for 12 h. The solution is added to ice water (800 mL) and the precipitate is filtered, washed with water and acetonitrile, and dried. To ethanol (600 mL) in an ice bath is added dropwise thionyl chloride (24 g). After stirring the solution for 15 min at room temperature, the precipitate is added, and the mixture is refluxed for 12 h. After the solvent is evaporated, chloroform (1000 mL) is added, and the solution is washed with 5% aqueous $NaHCO_3$. The separated organic layer is then dried over $MgSO_4$, filtered, concentrated, and purified by column chromatography on silica gel (eluent: CH_2Cl_2:MeOH = 99:1) to give the product as a white solid (5.01 g, 47.1%).

Compound **C** (diethyl-4'-(4-(4,4,5,5-tetramethyl-1,3,2-dioxaboryl)phenyl)-2,2':6',2''-terpyridine-6,6''-dicarboxylate) is prepared as follows. To 40 mL of DMSO solution of compound **B** (5.32 g, 10.0 mmol) are added bispinacolatodiboron (2.79 g, 11.0 mmol), potassium acetate (4.91 g, 50.0 mmol), and $PdCl_2(PPh_3)_2$ (702 mg, 1.0 mmol). The reaction mixture is stirred at 120 °C under nitrogen atmosphere for 12 h. After the resulting mixture is cooled at room temperature, the catalyst is removed by filtration and washed with chloroform. The filtrate is washed with H_2O. The separated organic layer is dried over $MgSO_4$, filtered, concentrated, and purified by column chromatography on silica gel (eluent: CH_2Cl_2) to give the product as a white solid (5.21 g, 89.9%).

© National Institute for Materials Science, Japan 2019
M. Higuchi, *Metallo-Supramolecular Polymers*, NIMS Monographs,
https://doi.org/10.1007/978-4-431-56891-9_11

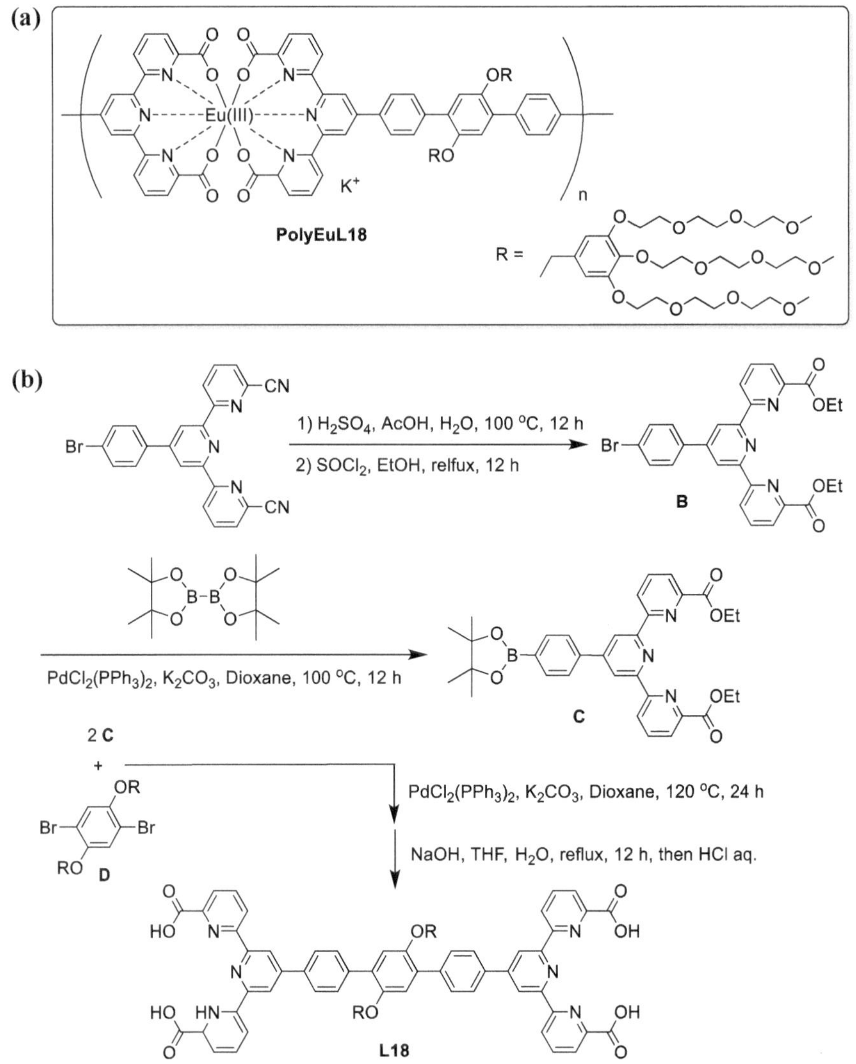

Fig. 11.1 a A Eu(III)-based metallo-supramolecular polymer (**polyEuL18**). **b** Synthesis of **L18**

L18 is prepared as follows. To 2 mL of DMSO solution of compound **C** are added compound **D** (2,5-bis(3,4,5-tris(2-(2-(2-methoxy-ethoxy)ethoxy)ethoxy)benzyloxy)-1,4-dibromobenzene) (142 mg, 0.10 mmol), potassium carbonate (69.1 mg, 0.50 mmol), and $PdCl_2(PPh_3)_2$ (7.02 mg, 10 mol). The reaction mixture is stirred at 120 °C under nitrogen atmosphere for 24 h. After the mixture is cooled to room temperature, the catalyst is removed by filtration and washed with chloroform. The filtrate is washed with H_2O. The separated organic layer is dried over $MgSO_4$, filtered, concentrated, and purified by column chromatography on silica gel (eluent: CH_2Cl_2) and preparative GPC to give the product as a viscous solid (38.1 mg, 8.8%). To a 5 mL of a THF-H_2O (v/v = 1:1) mixture solution of the viscous solid (170 mg, 79 µmol) was added sodium hydroxide (12.6 mg, 314 µmol). The solution was refluxed for 12 h, then was cooled at room temperature and neutralized with 1 M aqueous HCl. THF was removed under reduced pressure. The aqueous layer was extracted with chloroform. The separated organic layer was dried over $MgSO_4$, filtered, concentrated and purified by preparative GPC to give **L18** as yellowish-white solid (117 mg, 71.8%).

PolyEuL18 is obtained as follows. Equimolar amounts of **L18** and $Eu(NO_2)_3$ are refluxed in argon-saturated absolute methanol for 24 h. After the reaction mixture is cooled to room temperature, the solution is added to a mixture of *n*-hexane/chloroform (v/v = 1:1). The precipitate is filtered, washed with water and chloroform, and dried. The film is collected and dried further *in vacuo* overnight to give **polyEuL18** as a yellow solid (93.2%).

11.2 Vapoluminescence

A **polyEuL18** thin film is prepared on a glass substrate by spin-casting (20 µM in methanol). This film shows a bright-red emission under a UV lamp (λ_{ex} = 365 nm). The emission spectrum of a **polyEuL18** film shows the characteristic sharp peaks corresponding to the Eu(III)-centered $^5D_0 \rightarrow {}^7F_j$ (j = 1–4) emission bands between 550 and 720 nm, and the hypersensitive $^5D_0 \rightarrow {}^7F_2$ transitions (613 nm) dominate the spectrum. The ligand emission completely disappears in the polymer probably due to efficient energy transfer to the metal. The PL quantum yield (QY, Φ_{FL}) of **polyEuL18** in the solid state is 0.24, which is over ten times higher than that of **L18** (Φ_{FL} = 0.021). Interestingly, we found that the photoluminescence of the **polyEuL18** film can be switched off and on by exposure to acidic or basic vapor. When this polymer film is exposed to a HCl-gas-enriched environment for 5 s, the red luminescence disappears. After subsequent exposure of the non-luminescent film to triethylamine (Et_3N) for 5 s, the red luminescence is restored. Such behavior is called vapoluminescence. The **polyEuL18** film exposed to Et_3N vapor shows stable red emission during at least 100 h after the exposure. In contrast, the polymer film exposed to HCl gas shows the restoration of red luminescence after 48 h, suggesting that the acid is neutralized in the air. The response time is too fast to be measured. The quenching mechanism is unclear yet, but it is suggested that the protonation of the ligand in the polymer quenches the emission of Eu(III) ions (Fig. 11.2). Characters can be printed onto a glass substrate using a methanol solution of **polyEuL18**. The printed characters also show red emission under irradiation by a UV lamp (λ_{ex} = 365 nm), and the appearance and disappearance can be controlled using Et_3N/HCl vapors. Furthermore, the printed image maintains its luminescence for at least 4 weeks without any degradation.

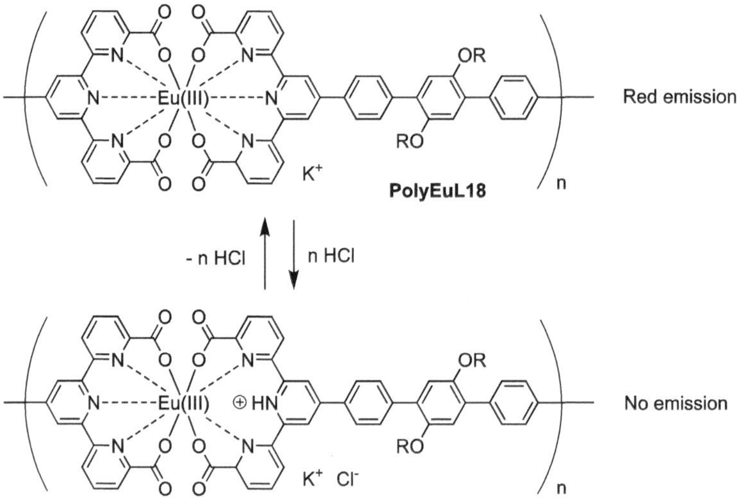

Fig. 11.2 Vapoluminescence in **polyEuL18** triggered by HCl vapor

11.3   Eu(III)/Fe(II)-Based Heterometallo-Supramolecular Polymer

The Eu(III)/Fe(II)-based heterometallo-supramolecular polymer (**polyEuFeL19**) is prepared by the complexation of Eu(III), Fe(II), and **L19** (Fig. 11.3) [2].

For the synthesis of a heterometallo-supramolecular polymer with Eu(III) and Fe(II) ions introduced alternately (**polyEuFeL19**), a new asymmetrical ditopic ligand (**L19**) is required.

L19 is prepared as follows. To a DMSO solution (23 mL) of 4′-(4-(4,4,5,5-tetramethyl-1,3,2-dioxaborolan-2-yl)phenyl)-2,2′:6′,2″-terpyridine (653 mg, 1.50 mmol) are added diethyl-4′-(4-bromophenyl)-2,2′:6′,2″-terpyridine-6,6″-dicarboxylate (799 mg, 1.50 mmol), potassium carbonate (442 mg, 3.20 mmol), and $PdCl_2(PPh_3)_2$ (105 mg, 0.150 mol). The solution is stirred at 120 °C under nitrogen atmosphere for 24 h. After the resulting mixture is cooled to room temperature, the catalyst is removed by filtration and washed with chloroform. The filtrate is washed with H_2O. The separated organic layer is dried over $MgSO_4$, filtered, concentrated, and purified by column chromatography on alumina (eluent: CH_2Cl_2) and preparative GPC to give a viscous solid of **L19** (940 mg, 82.4%).

The complexation behavior of **L19** with $Eu(NO_3)_3$ and $Fe(BF_4)_2$ was investigated in detail by UV–vis spectroscopy. The spectrum of a methanol solution of **L19** (20 μM), which is deprotonated with two equivalents of tridodecylamine, changed with the addition of $Eu(NO_3)_3$ up to a molar ratio of $[Eu(NO_3)_3]/[\mathbf{L19}] = 0.5$: the **L19** absorbance around 345 nm increased and the absorbance at 303 nm decreased. A plot of the absorbance at 345 nm vs the molar ratio of Eu(III) ions added to **L19** shows a linear relationship. Then, $Fe(BF_4)_2$ is added to the solution ($\mathbf{Eu(L19)_2}$) up

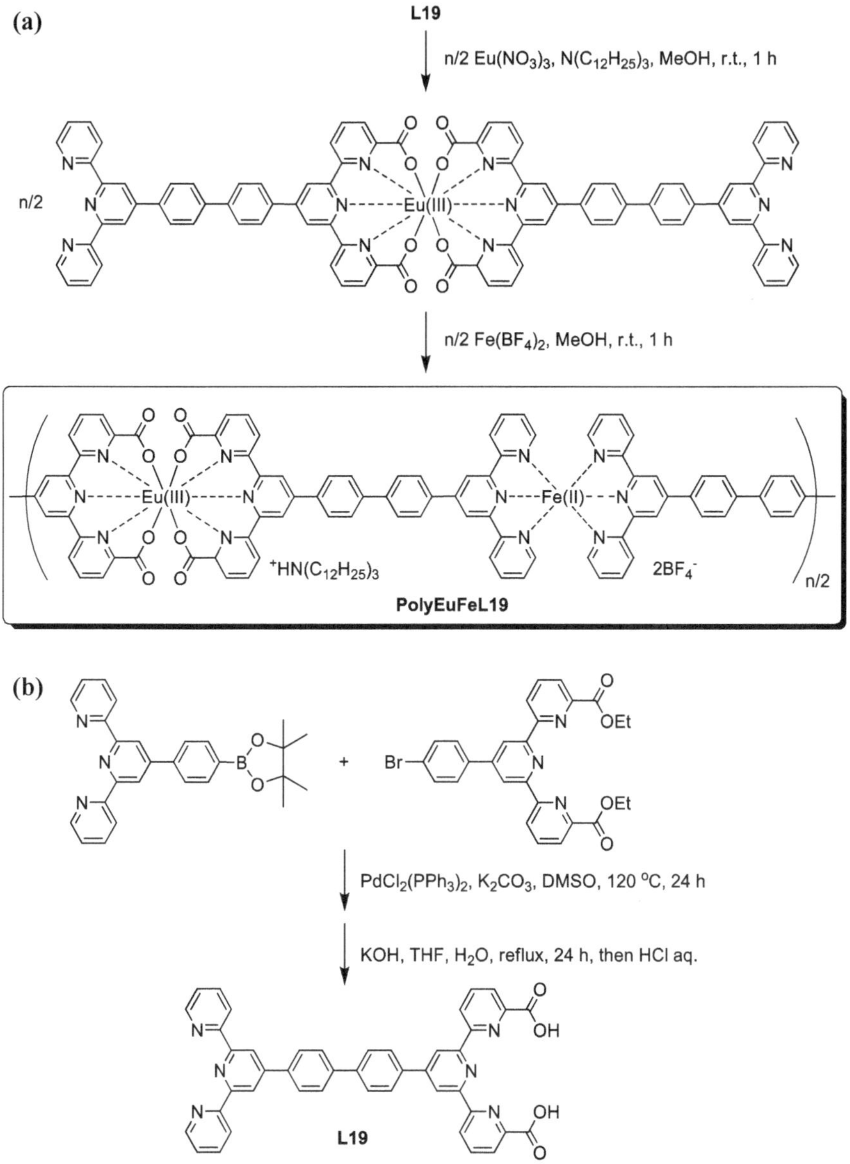

Fig. 11.3 Synthesis of **a** Eu(III)/Fe(II)-based heterometallo-supramolecular polymer (**polyEuFeL19**) and **b L19**

to a molar ratio of $[Fe(BF_4)_2]/[\textbf{L19}] = 1.5$. A new absorption at 570 nm based on the MLCT band appeared during the titration. The MLCT absorption was clearly saturated at a ratio of $[Fe(BF_4)_2]/[\textbf{L19}] = 0.5$. These spectral changes suggest the formation of a heterometallo-supramolecular polymer with Eu(III) and Fe(II) ions alternating in the structure (**polyEuFeL19**).

PolyEuFeL19 is obtained in a one-step synthesis by the complexation of **L19** with the addition of 0.5 equiv. of $Eu(NO_3)_3$ and the further addition of 0.5 equiv. of $Fe(BF_4)_2$ in the presence of tridodecylamine. It is a violet solid produced in 36% yield. It is soluble in alcohols such as methanol and ethylene glycol or protic polar solvents such as DMSO and DMF, but insoluble in common organic solvents such as n-hexane, chloroform, dichloromethane, and tetrahydrofuran. The molecular weight of **polyEuFeL19** was determined by SEC-Viscometry/RALLS using poly(ethylene glycol) as a standard ($M_w = 1.7 \times 10^4$, $M_w/M_n = 1.3$); this molecular weight strongly indicated that **polyEuFeL19** forms a polymeric structure. It is fully identified by its emission spectrum and cyclic voltammogram. The emission spectrum in a dilute solution of ethylene glycol exhibited characteristic peaks corresponding to the transitions from $^5D_0$ to $^7F_{1-4}$ in Eu(III) ions in the range from 550 to 720 nm, which is close to the wavelength of the MLCT band of the Fe(II)-tpy complex. The intensity of the emission peaks is very low (absolute quantum yield (Φ): 0.07) compared with that of **polyEuL18** because of the energy transfer from Eu(III) to the Fe(II)-tpy moiety. The intensity is also lower than that of the mixture of **polyEuL18** and Fe(II)-based metallo-supramolecular polymer (**polyFeL1**) ($\Phi = 0.12$), which indicates that the alternating complexation of Eu(III) and Fe(II) ions promoted effective quenching.

11.4   Emission Switching

The cyclic voltammogram of **polyEuFeL19** exhibits a reversible redox wave of the Fe(II)/(III) couple ($E_{1/2} = 0.8$ V). A thin film of the polymer was prepared on an ITO glass substrate by a solvent-casting method from an ethylene glycol solution (1 mg/mL) and dried in vacuo. The UV–vis spectra of the polymer film in an electrolyte solution of n-Bu$_4$NClO$_4$ (0.10 M) shows a reversible electrochromic behavior by the electrochemical redox of Fe ions: the MLCT absorption at 570 nm of the Fe(II)-tpy complex disappeared when Fe(II) was oxidized to Fe(III) at 2.0 V and reappeared when Fe(III) was reduced to Fe(II). This behavior is similar to that in **polyFeL1** and strongly supports the formation of the 1:2 complex of Fe(II) ions and tpy in **polyEuFeL19**. A **polyEuFeL19** film showed very weak photoluminescence (PL), but a remarkable enhancement of the red luminescence at 613 nm was observed by the oxidation of Fe(II) ions to Fe(III) at 2.0 V. The red luminescence was reversibly quenched by the reduction of Fe(III) to Fe(II). The excitation spectrum of the polymer ($\lambda_{em} = 613$ nm) indicates that **L19** acted as a photosensitizer. Therefore, the Eu emission was efficiently quenched by the energy transfer from Eu(III) to Fe(II) ion (the left energy diagram of Fig. 11.4). In contrast, when Fe(II) ions were oxidized to Fe(III), the Eu emission appeared, because the energy transfer from Eu(III) to

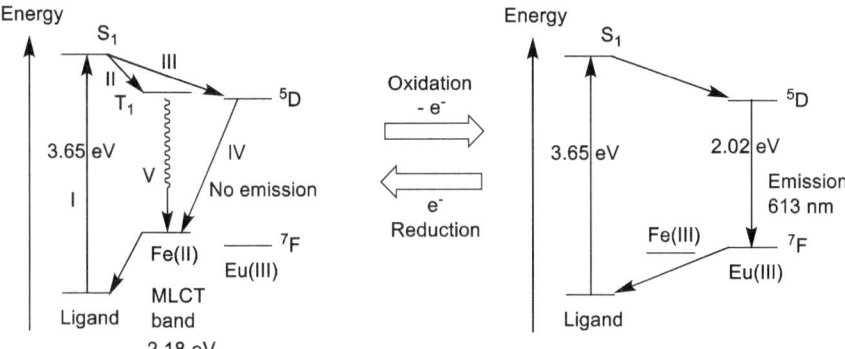

Fig. 11.4 Estimated energy diagrams of **polyEuFeL19** during the oxidation and reduction of Fe ions (I: absorption in ligand, II: intersystem crossing in ligand, III: energy transfer from ligand to Eu(III), IV: energy transfer from Eu(III) to Fe(II), IV: non-radiative transition in the MLCT band)

Fe(III) ions did not occur owing to the lowering of the HOMO level of the Fe ions, which was confirmed by the blue shift of the MLCT absorption band (the right energy diagram of Fig. 11.4). The repeated spectral changes in the polymer-coated ITO glass were recorded as a function of the applied potential. The luminescence intensity at 613 nm increased gradually upon incrementally increasing the potential, and the absorbance of the MLCT band at 570 nm decreased at the same time. The luminescence switching was reversible at least 10 times. Therefore, the polymer behaves as an "on-off" luminescence switch. The exchange of counter anions from to during the redox reaction was indicated by the change in the solubility of the polymer after the redox reaction, but the electrochromic and emission properties were unchanged during redox switching. This means that the anions and electrolyte do not affect the photophysical properties.

References

Sato T, Higuchi M (2012) A vapoluminescent Eu-based metallo-supramolecular polymer. Chem Commun 48:4947–4949. https://doi.org/10.1039/c2cc30972f

Sato T, Higuchi M (2013) An alternately introduced heterometallo-supramolecular polymer: synthesis and solid-state emission switching by electrochemical redox. Chem Commun 49:5256–5258. https://doi.org/10.1039/c3cc41176a

Chapter 12
Summary

This book has introduced the author's recent research on polymer design, synthetic methodologies, electrochemical/ionic/emissive properties, and the display device fabrication of Fe(II)-, Ru(II)-, Co(II)-, Cu(II)-, Pt(II)-, Ni(II)-, Cd(II)-, Mo(VI)-, and Eu(III)-based metallo-supramolecular polymers. The metallo-supramolecular polymers with a linear structure were prepared by the 1:1 complexation of the metal ions with ditopic ligands. As for polymer design, Fe(II)/Ru(II)-, Cu(I)/Fe(II)- and Eu(III)/Fe(II)-based heterometallo-supramolecular polymers were also described. The polymers were synthesized under the different complexation conditions for the two metal ion species or using an asymmetrical ligand. The metallo-supramolecular polymers with a hyperbranched structure were prepared using a tritopic ligand. The polymers showed various electrochemical/ionic/emissive properties including electrochromism, nonvolatile memory, and vapoluminescence. The Fe(II)-, Ru(II)-, Co(II)-, and Co(I)-based polymers exhibited blue, red, yellow, and black electrochromism, respectively. In addition, electrochromic display devices were successfully fabricated by combination with a gel electrolyte. The Fe(II)/Ru(II)- and Cu(I)/Fe(II)-based polymers exhibited multi-color electrochromism based on the different redox potentials between the metal ions. The ionic/protonic conduction in the Fe(II)-, Co(II)-, Ni(II)- and Mo(VI)-based polymer films at high humidity also proved interesting. A device with a polymer film served as a real-time humidity sensor. Regarding Ru(II)-, Zn(II)-, and Eu(III)-based polymers, unique emission properties were observed. The ON/OFF switching of emission was achieved with the Eu(III)/Fe(II)-based polymer.

Metallo-Supramolecular polymers are synthesized by the simple complexation of metal ions and multi-topic ligands. Therefore, the number of combinations of metal ion species and ligands is almost infinite. Not only the above-mentioned properties, but also other properties such as catalysts and magnetic behavior are expected to be found in these polymers. Actually, we are also investigating the biological applications of these polymers. The author hopes to arouse your interest in metallo-supramolecular polymers through this publication.

© National Institute for Materials Science, Japan 2019   89
M. Higuchi, *Metallo-Supramolecular Polymers*, NIMS Monographs,
https://doi.org/10.1007/978-4-431-56891-9_12